Life: Its Nature,
Origins and Distribution

JOSEPHINE MARQUAND M.A.
with a Foreword by
N. W. PIRIE F.R.S.

Life: Its Nature,
Origins and Distribution

The Norton Library
W · W · NORTON & COMPANY · INC ·
NEW YORK

First published in the Norton Library 1971
by arrangement with Oliver & Boyd Ltd.

SBN 393 00589 5

Books That Live
The Norton imprint on a book means that in the publisher's
estimation it is a book not for a single season but for the years.
W. W. Norton & Company, Inc.

PRINTED IN THE UNITED STATES OF AMERICA

1 2 3 4 5 6 7 8 9 0

To
MARY ADAMS

Preface

For over a century, biochemists have applied contemporary chemical and physical ideas in the intricate context of living systems. Their studies have revealed a remarkably high degree of organisation, interdependence and uniformity in all living and sub-vital systems on Earth today. These findings raise a number of questions as to the origins, antiquity and distribution of life, some of which are considered in this book.

I should like to thank N. W. Pirie, F.R.S., for his great kindness and generosity in making available his papers, research material, and for his advice at all stages in the preparation of the manuscript. I am also greatly indebted to Richard Marquand, Pearl Binder and Mrs Janka Dobai for their help and encouragement, and to Mr I. A. G. Le Bek and Miss O. M. Hamilton for their unfailing editorial assistance.

JOSEPHINE MARQUAND

Contents

Contents

Foreword

In spite of 300 years of experimental study, the nature and origins of life are still subjects for speculation. Increasing knowledge does, however, restrict the directions in which speculation can be taken seriously by biologists, and several phases in the process of mapping the confines of the problem can be recognised.

In the first phase the life-cycles of the smaller animals and plants were discovered, and it became reasonable to attribute all early observations on the 'spontaneous' appearance of visible organisms to the unsuspected presence of eggs, seeds or spores in the starting material. The second phase started during the 19th century, with the systematisation of knowledge about organisms visible only under the microscope. Putrefaction, fermentation and the agents causing many diseases were studied, and means of destroying or excluding them discovered. As a result, it became easy to set up vessels containing fluids in which organisms could grow, but which nevertheless remained unchanged until they had been seeded. This gave rise to the uneasy dogma that spontaneous generation was in all circumstances impossible and that organisms could not appear unless there had been contamination. Uneasy, because most of the scientists who were deeply concerned about the origins of life realised that the dogma must have been violated sometime, or somewhere. The dogma is still with us. Like all dogmas it inhibits research. It is obvious that recognised organisms do not appear if putrescible fluids are casually left in an incubator. It is dogmatic to assume that suitable sterile fluids, left in suitable conditions, can never undergo relevant processes of polymerisation and integration.

We are now in the third phase. In this, the biochemical

phase, knowledge is accumulating that may give the words 'suitable' and 'integration' some meaning. Hitherto, organisms have been recognised because they moved or brought about a transmissible change in some system exposed to them. We now make more intimate generalisations and say that all known organisms contain certain substances, or types of substance, and depend on certain enzyme actions. These statements may well be true, but the deduction that all organisms must contain or do these things does not follow logically. The organisms we know are the end product of a vast period of selection and extinction in one environment. They are the organisms that have been most successful in contending with each other and with the physical hazards of this environment, while depending on the chemical components that the Earth initially provided and that other organisms have made. The observed biochemical uniformity may do no more than give evidence about the completeness of this selective process. Evidence about the temporal distribution of life is not very helpful here. Fossil structures more than 2000 million years old are known that resemble present-day organisms; but if they had not borne this resemblance they would not have been recognised as fossils. It is certain that many other organisms will have left no trace and probable that some are not being recognised because they were chemically or morphologically bizarre.

The fourth phase will be ushered in when extensive study of the spatial distribution of life becomes possible. It seems certain that none of the higher forms of terrestrial life could survive in the open on any other member of the Solar System. Some of our bacteria could probably survive on some planets. It is completely fallacious to argue from this that the only possible forms of extraterrestrial life will be very simple. They may be complex – but different. It may be useful to consider in advance the three most obvious possible results of close biological study of samples from a planet such as Mars. No organisms may be found nor any evidence that there have ever been systems there that we would wish to call living. This will mean that the pre-conditions for biological integration are more stringent than many of us have thought. Something similar to terrestrial life may be found: not necessarily the 'bug-eyed monsters' beloved of the cartoonists, but metabolic systems depending on proteins, nucleic acids and other familiar types of molecules. This will be the least significant result, for it could be taken as evidence either that these molecules have a unique merit and that life is dependent absolutely on their properties, or that terrestrial and extraterrestrial life came from a common source. The most interesting result would be the discovery of organisms making use of chemical processes not used on Earth. If these are used exclusively, they will be strong, if not conclusive, evidence for independent origins. If the use of novel processes is extensive but not exclusive, it will not be possible to rule out a common origin followed by prolonged independent selection. This state of affairs will however contribute enormously to our understanding of the nature of life, that is to say, of the fundamental qualities, if any, that distinguish vitality from every other type of integration.

N. W. PIRIE

1. Is 'Living' a Useful Scientific Term?

And so they teach and babble undeterred –
With fools there's not a hope of intervening –
And when the people hear a sounding word,
They stand convinced that somewhere there's a meaning.*
Mephistopheles, GOETHE'S Faust

At every stage in the history of science there are some theorists who are not content to state simply that they do not know, but prefer to cloak ignorance with an essentially meaningless word or concept. One such concept is that of 'Vital Force'. Vitalists have generally maintained that inanimate nature is, as it were, a bottle into which vital essences (indefinable in terms of ordinary physics and chemistry) may be poured. In opposition to this is the view held by the majority of scientists, that though vital events are characterised by an exceptionally intricate deployment of physico-chemical processes, these processes are part of the stock-in-trade of the rest of science.

Isaac Newton, who compared himself to a child playing with pebbles on the shore of a vast ocean, wrote (1704): 'Transformations are very conformable in Nature.' After 1909, the year of the publication of Jean Perrin's well-known paper entitled '*Mouvement brownien et réalité moleculaire*', the regularity of 'transformations' began to be unequivocally interpreted in terms of precise and determinable molecular structures. The acceptance of the molecular theory resulted in a rapid expansion in the field of biochemistry, the study of the way in which substances act *in vivo* and in chemistry, their composition and structure.

Newton's foresight in the field of physiology was matched a

* Translated by Philip Wayne.

1

century later by that of the poet, Samuel Taylor Coleridge. Coleridge urged his contemporaries to 'take a wider view of life, which might fill the gap between physics and physiology, and justify using the former as means of insight into the latter'. Coleridge's exhortations were premature. They are less so today. J. B. S. Haldane, who for 35 years wrote about the nature and origins of life, and proposed at least one novel idea in every article he published, often applied quantum mechanical theory to biology. Like Coleridge, but unlike many biologists, Haldane was aware of the value of exploring the biological consequences that would be entailed if the conclusions being reached in physics proved to be well founded.

Are we, then, within reach of a rigid operational definition of life? The 'fundamentals' of a subject are often the least secure part of it. We have an aesthetic appreciation of the phenomenon of 'life' and cannot dogmatically state that there is no underlying physical reality – which is clearly what the original users of the word 'living' had in mind when they distinguished cats and rabbits from water and stones. But our experience is now wider. More intermediate materials are known. The consideration of systems such as the viruses, metabolically active cell or tissue fragments, or life-like systems that may operate, or may have operated, under environmental conditions different from those on Earth today, serves to emphasise the uncertainty that exists when we say that a system is or is not alive.

A dog is clearly alive. In an environment provided with suitable quantities of food, oxygen and water, it is irritable, motile, fecund and so on. By anaesthesia, sleep or castration any of these qualities may disappear without the dog being dead. If its head is cut off, the dog, considered as a dog, is dead. Its tissues are not. A physiologist working with one of its kidneys on a perfusion pump will consider the kidney alive if it can continue to respire and secrete. If secretion fails, the physiologist would maintain that the kidney had died. In a suitable culture medium, however, a histologist might still be able to culture individual cells from this kidney so that they could grow and divide.

Scientists have devised a number of criteria to explain what they mean when they claim that a system is 'alive'. Traditionally, it was held that a system was alive if it moved. This criterion was found wanting. It was not really applicable to eggs or plants and to an increasingly large number of other systems that seemed more living than dead. Later criteria have included the manifestation of metabolism, a certain chemical composition, cell structure, a minimum size, the manifestation of optical activity and isotopic fractionation, and the ability to replicate. Each new criterion marked an advance in our knowledge of biological processes and in our awareness of their relationship with the physical sciences.

Each advance also emphasised that each criterion, though well adapted to certain organisms, was ill adapted to others. What is more, each can be shown by a non-living or a man-made system (see Chapter 2). A number of multicellular organisms – worker bees and brain cells, for example – do not reproduce. A number of organisms have stages in their life-cycles during which they 'play dead' for prolonged periods of time, suspending all metabolic activity until external conditions have become sufficiently favourable for them to resume metabolic activity. (The resistant phase in the life-cycle of many micro-organisms caused a great deal of confusion when metabolic criteria first came into use in the 19th century, as we shall see.)

On the other hand, reproduction or the manifestation of some form of metabolism can be demonstrated in a number of non-living systems, such as flames, crystals, cloud patterns, fern-like inclusions in agate, viruses and genes. We could call these systems 'non-vital analogues'. Even though they may not satisfy our minimum (though generally unspecified) standards for organisms, they help us to interpret more complex interactions characteristic of biological systems (see Chapter 2).

By the first decade of the 20th century, a number of scientists had come round to the view that it was not going to be possible to find any single criterion that could be used as a touchstone for life which was not simultaneously applicable to what most people would agree was a non-vital system as well. E. A. Schafer pointed out, in his presidential address to the British Association (1911), that the only universally acceptable definition for life seemed to be on the lines of the cynical circular definition of an archdeacon as 'a man who performs archediaconal functions'. Schafer was speaking at the inauguration of a stimulating period in the application of molecular theory to biology. It was already apparent that though the plain man might think he knew what he meant when he said that a system was alive, the scientist did not when he spoke of the systems, such as enzymes, that most interested him when he attempted to analyse life.

CRITERIA THAT HAVE BEEN USED TO IDENTIFY VITAL SYSTEMS

The most ancient and readily apparent criterion for life was that of movement. This is a legacy from language. The Bible contrasts the 'quick' with the 'dead', and scientists still use the word 'viviparous' to describe those animals that produce active young. A similar linguistic relic is to be found in the word used by the original inhabitants of Australia to describe reaches of water that are not flowing: a bilabong, 'dead water', is invested with the same stillness as invests a corpse.

The 17th-century Dutch microscopist, Anton van Leeuwenhoek, the first scientist to describe most of the groups of bacteria we recognise today, used the criterion of movement systematically in all his descriptions of micro-organisms. It was on the basis of this criterion that he was able to contest William Harvey's postulate that the foetus developed from an egg. It seemed obvious to Leeuwenhoek that fertilisation depended on sperms, which he could see wriggling under the microscope, and that the mother merely served as an incubator.

The use of the criterion of movement did not deter Leeuwenhoek, as a sound empiricist, from figuring static cocci among the bacteria he described. By the 19th century, many scientists realised that not all microscopic objects that were motile were also alive. In 1828, the botanist, Robert Brown, observed that grains of Lycopodium pollen suspended in a water droplet were agitated with great rapidity when viewed under the microscope. He inferred that this agitation must be a consequence of the movement, or, as we should express it today, the 'bombardment', of pollen grains by the water molecules in which the grains were suspended.

Two years before Robert Brown's perceptive inference was made, the Swedish chemist, J. J. Berzelius, had drawn a distinction between the large molecules derived from living organisms, which he termed 'organic', and the molecules derived from non-living systems, which he termed 'inorganic'. This sensible inference was drawn from incomplete knowledge. The distinction is still preserved, even though we are surprised today how few of the complex molecules known to synthetic chemistry are actually responsible for the structure and function of organisms. The continuity between what Berzelius called organic and inorganic chemistry had already been affirmed half a century earlier by the French chemist, Lavoisier, who had claimed that no distinction should be drawn between the elements present in living and non-living systems. Alcohol, a 'plant' product, had been synthesised. In 1828 one of Berzelius's ablest pupils, the chemist Friederich Wohler, wrote to his former teacher: 'I must tell you that I can prepare urea without extracting it either from the kidney of a man or a dog.' Wohler had proved Lavoisier's thesis at the molecular level, by converting the inorganic chemical, ammonium cyanate, into its 'organic' isomer (see Chapter 6), the characteristic 'animal' product, urea.

Berzelius, Wohler and another former pupil of Berzelius, Justus von Liebig, became staunch proponents of the view that the metabolic processes of vital systems should be interpreted in terms of ordinary chemistry. This view brought them into the controversy that had raged since the 18th century as to the vital status of the yeasts and moulds. Yeasts and moulds had been known in the kitchen since ancient times. They did not move, they appeared spontaneously, as far as people knew (see Chapter 3), and the suggestion that they were organisms was, according to Coleridge, viewed with as much 'contemptuous surprise' as the old idea that metals were alive (see Chapter 2).

In 1826 and 1827 two biologists, Cagniard de La Tour and Theodor Schwann, independently claimed that the non-motile, globular particles of yeast that could be examined under the microscope were sufficiently organised to qualify as organisms. Schwann pursued this argument further, claiming that 'the ferment is generated and increased by the process of fermentation, a phenomenon only met with in living systems'. Berzelius had done a great deal of work on what he termed 'catalysts' (substances which accelerate chemical changes, like the fermentation of sugar to alcohol, without themselves being altered in composition). He asserted in 1839: 'Yeast is a mere catalyst. It is no more living than a precipitate of alum.' *

Using the pseudonym 'S. C. H. Windler', Wohler and von Liebig lampooned the biologists' view, describing yeasts as: 'Albuminoid animalcules without teeth or eyes, which devour sugar. From their anus issues a stream of alcohol, which rises to the surface, and from their genitals comes carbon dioxide.'

Although the chemists were justified in pointing out that the biologists' view tended to ignore the chemical basis of fermentation, they had a crude view of the ways of micro-organisms. A decade later, Schwann, in collaboration with Matthias Schleiden, formulated the cell theory, which maintains that the cell provides the basis for all plant and animal life. This view quickly found acceptance. It was clear that great advantages flow from the ordered juxtaposition of the contents of a cell in a semi-permeable membrane, which simultaneously protects them and keeps them in contact with the external environment.

Cells were soon shown to have a very wide size range. They could be as large as those in the sciatic nerve of a horse, or as minute as the ones found in the smallest bacteria. One group, below the size of what were usually classified as Bacteria, was found at the very end of the 19th century to be filterable

* One of the catalysts responsible for alcoholic fermentation was isolated at the end of the 19th century, and called 'zymase'. This term is the origin of the word 'enzyme', which is used to describe all biochemical catalysts. Each enzyme catalyses a specific reaction in the stepwise series of chemical exchanges known as metabolic pathways.

through a porcelain filter. This established an arbitrary though practical limit to the group classified as the viruses. The extreme parasitism of the viruses caused many scientists to infer that a virus, with a diameter of, say, 20 nm,* was too small to be able to carry on the complex activities of life. Then a few members of an ill-defined group known as the Myco-plasmataceae were isolated. Some of them cause infectious diseases like Pleuropneumonia in birds and cattle. They are the 'Pleuropneumonia-like organisms', or PPLO, and have a limiting diameter of about 125 nm. This diameter is well within the size range of some recognised viruses, but since some PPLO, unlike viruses, are free-living for parts of their life-cycles, they are recognised as the smallest free-living cells or organisms known today, just as the yeasts were 100 years ago.

In fact, very few living systems, except those bacteria and plants able to grow and multiply in a medium containing simple molecules, are able to grow and multiply without a supply of nutrients provided by other organisms. Some obligate parasites, like the Protozoan responsible for the trans-mission of malaria by mosquitoes and humans, rely on their host for some of the enzymes essential for their growth and multiplication. The smaller viruses seem to have no action on their hosts other than misdirecting the synthetic mechanisms of parts of the cell they have invaded. Many of them have no known independent enzyme activity, though a few carry enzymes that are involved in the initial attachment to the host cell.

Nevertheless, we can draw an approximate analogy between the extreme dependence of the viruses on the synthetic mechan-ism of a host cell and the dependence of many higher animals on other organisms within them. It could be argued, for in-stance, that the micro-organisms that make cellulose-decomposing enzymes in the cow's rumen are an essential prerequisite for the growth and multiplication of the cow. If these organisms could not digest cellulose, the cow would be unable to use grass as its principal foodstuff. We recognise by other criteria that a cow is living. But if the criteria we use for

* nm = nanometre = 10^{-9} metre.

borderline cases turn out to be different from those used for
obviously living or non-living systems, is there not something
fundamentally unsatisfactory about our definition?

Chemical composition and a high degree of organisation as criteria for life

'Windler's' lampoon on the biological theory of fermentation
described the yeasts as 'Albuminoid'. This term was used by
many 19th-century scientists to describe what Berzelius called
'protein', a substance that had been isolated, for example, in
egg yolk. (Proteins were known by the middle of the 19th
century to manifest the phenomenon of optical activity, as we
shall see in the next section.) Berzelius derived the word 'protein' from a liturgical reference to Adam as 'Protoplastus', the
first-formed body, basis of all subsequent life. Nineteenth-
century scientists were impressed to find that all forms of life
they had examined were made, among other things, of protein. Louis Pasteur, as able a chemist as he was microbiologist,
was expressing the ideas of his time when he wrote (1860):
'The presence of protein is indispensable to all forms of fermentation, for all ferments need protein to live.' This view was
echoed by Engels.

No one would deny that proteins offer the most effective
way of carrying on the affairs – enzymic or otherwise – of an
organism. But if we found a system lacking protein, but otherwise satisfying our requirements for life, would we deny it the
title 'living'? We cannot assume that substances found in all
organisms so far examined, such as proteins, nucleic acids,
phosphate esters, fats and carbohydrates, were equally important in earlier phases of biochemical evolution, or in
environments where different chemical elements may possibly
form the basis for alternative forms of 'life' (see Chapter 7).

Optical activity as a criterion for life

Ordinary light vibrates in all directions at right angles to the
direction of transmission. When passed through certain
crystals at a suitable angle, some of these directions of vibra-
tion are partly or completely suppressed. The resulting light is

called polarised light. Such light will be transmitted by another suitably cut crystal having a similar orientation to the first. It will be stopped by one at right angles to this. It was observed at the beginning of the 19th century, when polarised light was first investigated, that various solutions rotate the apparent plane of polarisation of incident polarised light, a phenomenon commonly referred to as 'optical activity'. The optically active solutions investigated were invariably of biological origin. This caused Pasteur (1870) to suggest, after he had studied optical activity and the problems of inducing the phenomenon artificially in inactive substances, that: 'Optical activity is perhaps the only sharply defined boundary which can be drawn at the present time between the chemistry of dead and that of living matter.' (See Chapter 6.)

Most present-day writers give optical activity a prominent place among vital criteria, and would probably accept as evidence for its vital origin the presence of a high degree of optical activity in a sample of material found on Earth or coming from an extraterrestrial source. Although they contain optically active molecules, most organisms have a very limited capacity for making them from inorganic materials. Most of the metabolism of a mammal, for example, consists in rearranging optically active molecules derived from the organisms on which they feed. It is thus the habitual use of optically active molecules, not their creation, that is characteristic of an obviously living system. Although we might say there is nothing surprising in optical activity in organisms, is it peculiar to them?

Segregated areas in naturally occurring deposits of minerals such as quartz are sometimes found that have a slight preponderance of one optically active form. It is possible, because of this, that complex organic flocculates or solutions could acquire a small degree of optical activity by preferential absorption on to these active surfaces (see Chapter 6). Optical activity is known to survive some rock-forming processes. For example, some petroleum fractions are optically active. Suppose a sample of oil has percolated through great thicknesses of rock-forming minerals with a slight preponderance of one optical type. The optical activity of such a sample may be due to its vital origin, but it cannot be assumed *a priori* that this is so, for the activity might also have arisen as a result of differential absorption on the rocks through which the oil has percolated.

Isotopic fractionation as a criterion for life

Living systems, like other systems characterised by a series of stages of absorption, diffusion and phase equilibrium, can differentiate between the isotopes of many of the biologically important elements. This characteristic has not been put forward as a criterion for vitality, for its existence was not appreciated until people had already begun to doubt the value of trying to define vital systems.

Living organisms pack into a relatively small compass the sort of capacity for isotope fractionation that would demand a long separation column in an isotope factory. There are two stable isotopes of carbon, with atomic weights 12 and 13. It is characteristic of living organisms that they contain a larger proportion of C-12 than the average non-living substance, in which most of the world's carbon is present. The existence of an unusual isotope ratio is *prima facie* evidence that a system is or was living, but it is not conclusive evidence. Given time and distance, isotope fractionation proceeds wherever absorption, diffusion or a phase change establishes a set of equilibria. Moreover, fractionation is not systematic in organisms. Thus the distribution of carbon isotopes in different parts of a tree is variable. The presence of proteins or other optically active molecules, or the existence of isotopic fractionation in a system, cannot therefore be taken as evidence that it will satisfy all the implications of the word 'living', nor should their absence be used as final evidence that it will not.

Aesthetic criteria for life

Even if we cannot frame a definition of the term 'life', is it possible to describe how to recognise systems which are best described as 'living'?

William Blake wrote (1810): 'I question not my Corporeal

or Vegetative Eye any more than I would Question a Window concerning a Sight: I look through it and not with it.' This attitude presupposes an underlying reality either in the external world, or in the head, or in both, about which the facts give some information. Interpretations are part of the innate or early acquired fabric of our sense impressions. Many of them are misleading, as Aristotle's trick of rubbing one's nose with crossed fingers shows. We are familiar with the vital form and think we can recognise it when we come across it. Unfortunately, aesthetic criteria are valid only when we are dealing with vital systems showing a very elaborate structure and metabolic activity. It is when we are dealing with borderline cases, on the boundary between non-vital and vital, that it may be impossible, using aesthetic criteria alone, to differentiate between a non-vital analogue to life and a living system.

We can make a crude model of an amoeba cell from protein, gelatin and the carbohydrate, gum arabic, which may appear to imitate the 'streaming' movement of protoplasm. Artificial algae and fungi have been constructed, but their resemblance to organisms is irrelevant to an understanding of the nature, if any, of life. Although the rhythmic banded precipitations of inorganic chemicals studied by Leisgang (which are still inadequately explained, but which appear to be able to integrate themselves without vital intervention) might appear to our unaided senses to be of vital origin, we know that they are not. They lack the very elaborate structure which enables us to recognise organisms unequivocally. On the other hand, we consider it probable for similar reasons that certain deposits with a regularly disposed lamellar structure found in the rocks of ancient shields are of vital origin. It is partly on account of such lamellar deposits that the most probable time for the origins of life appears to pre-date the middle Pre-Cambrian. It seems capricious to try to interpret these structures in any non-biological way.

The similar attempt to assign a vital origin to certain structures in meteorites is not generally accepted. The small class of stony meteorites contains on the average about 2% of carbonaceous material. Many of these meteorites have been examined for vital remains by numerous investigators since Berzelius's time. Most of the structures found in them have been dismissed as contaminants. This was Pasteur's view after a careful examination of some of the material available in his day. Nevertheless, a few structures present in carbonaceous meteorites continue to puzzle some scientists. Such structures have been variously described as 'lenticles', 'vesicles', 'involutions', 'pellicles' or 'pseudomorphs'. To establish their possible biological origin, new techniques and criteria are needed. Some investigators circumvent the problem by assuming that these apparently borderline and much-altered organic structures are the primitive and metamorphosed stages in the generation of life from inorganic precursors. However possible this may be, it is apparently no easy matter, using aesthetic criteria, to establish the boundary between living and non-living in those cases where satisfactory criteria might be most valuable.

The terms 'living' and 'non-living' are still useful for many purposes, in the same way that the words 'green' and 'yellow' are useful. So far as our unaided vision is concerned, it is obvious that there is a smooth transition between the two colours. As relatively little of our colour experience falls into the region 'greenish-yellow', we name regions of the spectrum as we do for convenience, and imagine they touch upon legally definable lines. Colour distinctions can be defined in terms of a single quantity; life cannot. Though they might be drawn up to exclude all obviously non-living systems, any combination of two or three qualities will also exclude some systems which are, if not typically living, at least generally included in that category. Confusion does not often arise, as most research is carried out in regions far removed from the boundary. Most systems are either obviously living or obviously dead. When the matter becomes doubtful, Ritchie's neat phrase is generally held to summarise the matter: 'Some things are deader than others, some things are livelier than others.'

This attitude to life, which might be labelled Restrained Induction, or, less flatteringly, Empirical Nihilism, has not gained universal acceptance. A surprisingly large proportion of scientists at international conferences get bogged down from time to time in metaphysics about the nature of life. They cannot define nor unequivocally recognise living systems, but they are none the less certain that such entities could be defined. The history of science offers many examples of useful discussions on ill-defined or undefinable themes. It is when the status of such themes is thought to be established that a discussion of them may become unprofitable. It would therefore seem prudent to refrain from saying that certain observations on a system have proved that it is or is not 'alive', and to avoid the use of the word 'life' or 'organism' in any discussion of borderline systems.

2. Analogies

'When we no longer look at each organic being as a savage looks at a ship, as something wholly beyond our comprehension; when we contemplate life as the summing up of many contrivances, each useful to its possessor, in the same way as a great mechanical invention is the summing up of the labour, the experience, the reason, and even the blunders of numerous workmen, how far more interesting – I speak from experience – does the study of natural history become.'

CHARLES DARWIN, 1859

Interpretations of nature are comprehensible in proportion to their simplicity. The simpler a scientist can make his formulations, the more fully other scientists will understand him, and the larger the number of inexpert who will understand him partially. It is out of this widespread understanding that the next stage of progress comes. One of the principal ways in which scientists from the earliest times have attempted to clarify phenomena near the limits of their understanding has been by the use of analogy. This device is useful, but tends to mislead its users into thinking they know more than they do about a subject.

Analogies always have a contemporary flavour. They share the defects of slang. Though vivid, they can be misleading in a serious statement at any time, and become particularly so when they are out of date. One scientific fallacy that has long been perpetuated by false analogy is the vivid notion, dating at least from the 18th century, of the 'crystallisation of life' from non-living matter. Vitalism had not been discredited when this analogy became fashionable. In its oversimplified form it lingered on (as we shall see in the second half of this book), encouraging people to think that micro-organisms of various kinds, or their components, arose as spontaneously as crystals from a suitably seeded saturated solution. This view

14

was opposed with vigour by evolutionists like Herbert Spencer and T. H. Huxley, who thought that organisms evolved very gradually over an extremely long period of time.

René Descartes (1596–1650), in an age when very little was known about the structure of matter, thought of life in terms of a machine operated by interlocking splints and spheres, which kept in constant motion until its mechanism wound down. He postulated (1640) that this 'beast machine' (see also Chapter 5) could not be logically distinguished from a living organism. Mechanist interpretations of life were viewed with repugnance by Vitalists, but the large majority of scientists until the 19th century were prepared to compromise in their own minds between the Vitalist and Mechanist schools.

The compromise between Vitalism and Mechanism was already being made in Descartes's day. In 1630 Jean Rey wrote a treatise in which he attempted to refute the old notion that metallic ores were (like oysters) alive and capable of replacing themselves. If metals were capable of gaining weight because of growth, he claimed, how surprising it was that their 'soul' was able to survive the rigorous heating of purification. He tempered this Vitalist argument with a piece of sound empiricism, claiming that lead could not be alive 'because it was a homogenous body without distinction of parts'. The biochemist, Frederick Gowland Hopkins, used a similar criterion for life when he stated (1913) that a minimum requirement for a vital system was 'a dynamic equilibrium in a polyphasic state'. (Some further minimum requirements for vital systems are discussed in Chapter 4.)

Two German physiologists in the 1870s were not prepared to restrict their view of vital systems to the solid/liquid phase boundary. Inspired by the incandescent theory of the origin of the Earth (see Chapter 4), Preyer and Pflüger envisaged the possibility of gaseous organisms in próbiotic times. Preyer thought that these organisms might have subsisted on a diet of meteorites, a suggestion which provoked the scorn of the Belgian botanist, Louis Errera. Pflüger's suggestions were less Wagnerian, and are current in modern discussions on the origins of life. He envisaged the probiotic origin of organisms in terms of the polymerisation (see Chapter 6) of heat-stable carbon–nitrogen derivatives known as 'cyanogen' compounds, and deserves great credit for recognising that constructive chemical processes were as possible as destructive ones in such high-temperature systems. If anyone today found a high-temperature system behaving in an even more life-like manner than flames do, the phenomenon would be called an interesting analogy with life, rather than the beginning of Asbestobiology.

We will now examine some aspects of the flame analogy, which is a remarkably good and close one, and is, in a scientific context, at least as old as Heraclitus.

A flame can move over a field of dry grass catalysing the reaction between the oxygen of the air and the carbohydrates in the grass, leaving an unconsumable residue behind it. These words also describe what a rabbit does when it feeds in a field. A flame issuing from a fixed pipe depends for its existence on the oxygen of the air. It can be compared with a plant (which cannot move but which feeds on the material present in its environment). Life, like a flame, exists at the expense of a pre-existing environment. The essential thing is that this environment should be metastable, containing substances ready to react with one another, but unable to do so until the action is catalysed. Every system that could conceivably be called 'living' exists in an actual or potential energy flux. Its existence is entirely dependent on that flux. Green plants exist at the expense of the energy flux given by the light of the Sun, and this group is responsible for establishing the metastable states on which almost all other life depends.

Another striking peculiarity of organisms, besides their energy relations, is their capacity to reproduce. This capacity can be demonstrated in a large number of simpler systems such as flames, magnets, crystals, polymers, enzymes, genes and viruses. Among the replicating systems present in a 'living' cell, we can consider enzymes, since something is known about how some of them are formed. An enzyme 'precursor' such as trypsinogen or pepsinogen can form the end product, trypsin or pepsin, only when a trace of trypsin or pepsin is present in the medium in which replication is taking place. In this way, more enzyme is produced. A simpler system, manifesting the same characteristics, is the reduction of cupric to cuprous ions by means of molecular hydrogen. There is a prolonged time-lag unless a trace or 'starter' of cuprous ions is added to the system.

The essential sequence of events in these replicating systems is this: one or more particles at least partly resembling the end product is introduced into a suitable medium under precise environmental conditions. Replicas of the original structure are then produced from the components of the medium. Penrose and Penrose have illustrated some aspects of this process, which is characteristic of complex biochemical systems, with a few units of very simple character called 'parabionts'. These are pieces of plywood or vulcanite which are designed to link together, like pieces of a jigsaw, to form the double structures AB or BA (see Fig. 1).

Unlinked pieces A and B are scattered at random on a grooved and restricting track. When the track is shaken by the experimenter, the pieces A and B are able to slide horizontally without passing one another (Fig. 1 (a)). Replication is initiated by the experimenter introducing the linked structure BA (known as the 'seed') on to the track (Fig. 1 (b)). If sufficient oscillatory energy is provided and sufficient units are available, the seed 'activates' suitably orientated pieces with which it comes in contact, producing a set of linked units identical with itself ((c) and (d)). The narrow track, which restricts the units and facilitates replication, can be compared with a catalyst. The vital phenomena of mutation and chiral selectively are simulated (see Chapter 6) if the opposite form of 'seed', that is to say the parabiont AB, is introduced into the system (e). In these circumstances, parabionts of the form AB will be replicated exclusively, and the system can be said to have mutated or to have become resolved in an alternative form ((f) and (g)).

In this replicative model, the initiator and the end product are similar. In others they are not. The high molecular polymers known as polysaccharides are initiated in solutions containing suitable enzymes and a sugar phosphate, not by a

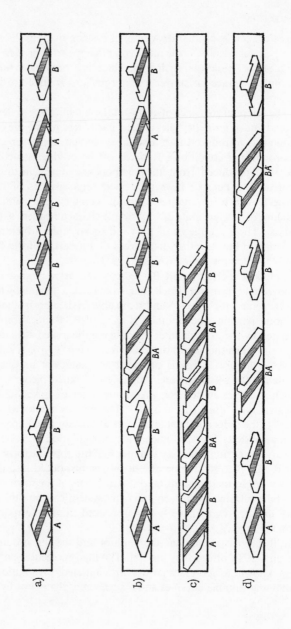

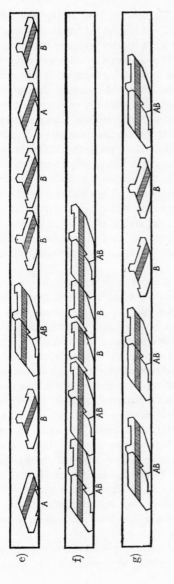

Fig. 1. Self-replicating machine made with units of two kinds. (a) Six units are shown, arranged, supposedly at random, on a track which is subject to horizontal agitation: the units do not link. (b) Seed, BA, formed by two linked units, is introduced. (c) Horizontal agitation now causes transmission of activation from the seed to adjacent units. (d) As the agitation gradually ceases, the units separate but the original seed, BA, remains intact and a second linked pair has been generated. (e) The situation here is the same as in (b) but a different seed, AB, has been introduced. (f) A different type of activation is produced on agitation. (g) After separation, two new pairs like AB are seen to have been generated. (Reproduced by kind permission of Professor L. S. Penrose.)

e)

f)

g)

trace of polysaccharide, but by one of the low molecular weight, unpolymerised saccharides known as oligosaccharides.

It is not necessary for a 'starter' to carry the full specifications for the replicative processes it initiates, any more than the coin that activates an automatic telephone is expected to design and assemble the telephone exchange through which a message is transmitted. This point can be illustrated by means of the crystal analogy. A crystal 'nucleus' does not carry within it the specifications for the construction and arrangement of the molecules from which the crystal is built. On the contrary, when a crystal nucleus is introduced into an environment that contains, or can make, the necessary molecules, the nucleus is offered various molecules in all orientations by the random operation of Brownian movement. If the material and its orientation are right, the crystal grows. The impulse for this growth comes, not from any kind of searching or organising capacity in the nucleus, but from the maintained chaos of the solution.

Another system that operates by means of the use to which it can put random events occurring in its environment has been described by W. Ross Ashby (1952) in the stimulating analogy he drew between a mechanical chess player (designed to compete at chess with its designer) and the process of evolution in biological systems. A mechanical chess player is designed to select from the various moves open to it the ones that will lead to the most successful result. In this way, it is able to outplay its designer, acting like a group of mutating organisms living in a somewhat hostile environment. The options offered by the environment are crucial to the direction in which the evolution of a structure or substance can proceed. The first steps of evolution may seem useless. They wait, Micawber-like, for a function to turn up. Hence the validity of Haldane's recipe (1954) for the evolution of life: 'Take a planet with some carbon and oxygen: irradiate it with sunshine and cosmic rays and leave it alone for a few hundred million years.' (See Chapter 4.)

All analogies have a contemporary flavour. In Descartes's day, life, which was held to be characterised by motion, was compared with the ticking of a clock. In Berzelius's day, the energy exchanges of an internal-combustion engine provided an illuminating comparison for the metabolic criteria that were just becoming fashionable. Today evolutionary processes, such as replication and copying with or without error, are compared with the organisation of random sources of information in an electronic computer. This analogy does not imply that life is the product of the activities of a feeble-minded coding clerk. Implicit in the analogies of any period will be some assumptions that will mislead and others which will point directions for future research. Analogies, when drawn to illuminate observations in the field of biology, are valuable only if we keep in mind suggestions coming from various directions. Only thus are the first glimmerings of fresh and relevant observations likely to be recognised and cherished.

3. On the Spontaneous Appearance of Life

'The secrets of Nature reveal themselves more readily when tormented by art than when left to their own courses.'

FRANCIS BACON, 1620

The ancients imagined that organisms like maggots, flies, bees, moths, eels, frogs, mice and even crocodiles could emerge spontaneously from non-living media like hay, mud, corpses or the refuse of their own backyards. Beliefs in the spontaneous generation of organisms were not confined to Greek and Roman tradition. The traditional saying, 'decaying grass becomes the firefly', is still in use at political meetings in China, and the 12th-century neo-Confucian philosopher, Chu-hsi, believed in the spontaneous generation of body lice. This belief was maintained as late as 1683 by the eminent chemist, Van Helmont, who wrote: 'Lice, bugs, fleas and worms become our guests [a phrase similar to that used by Chu-hsi] and are, as it were, born of our inward parts and our excrements.'

Close attention to experimental detail characterises those recipes for the generation of life that are still on record. The media chosen were by no means ones we should now describe as sterile. Virgil details with precision the measures which the shepherd Aristes took to produce a swarm of bees from the sides of an immolated bullock (*Georgics*, V). A Judaic variant of this experiment is recorded in the Book of Judges, where a swarm of bees emerges from the sides of a young lion dismembered by Samson.

This experiment was referred to with some vigour by Florentine academicians in 1668, when the Tuscan naturalist, Francisco Redi, published a carefully documented treatise claiming that he 'did not hold it a very grave matter' to deny the possibility of spontaneous generation. Redi suggested that the appearance of fly maggots on putrefying flesh was caused by adult flies laying their eggs there, and was able at one and the same time to open the attack on spontaneous generation and to invent the meat-safe. Redi used the finest Neapolitan muslin to exclude flies from vessels containing various different raw meats. He left his control vessels uncovered. Flies' eggs appeared in the muslin of the vessels he had protected. Eggs, and then maggots, appeared in flesh in the control vessels. Filtration through cotton was revived by 19th-century microbiologists who tried to exclude microscopic vital agents from nutrient media, as we shall see.

A number of scientists continued to claim that they had observed experiments in which large organisms had been spontaneously generated. Van Helmont claimed that he had produced mice from bran and the exhalations of a dirty shirt. James Harrington, social philosopher and friend of Charles I, built a rotating summer-house outside St James's Park, and shut himself up in it in order to observe the generation of flies from his own sweat.

Nevertheless, a number of scientists were beginning to suspect that many of the claims that were still being made for spontaneous generation were the result of the transference of pre-existing life from one manifestation to another. The epigram chosen by the designer of the frontispiece of William Harvey's treatise on the generation of animals (1680) expresses this view succinctly: '*Omne vivum ex ovo*' ('All life derives from the egg'). Leeuwenhoek's qualification of this view, that in larger animals spermatozoa were also necessary for fertilisation, was mentioned in Chapter 1. By the end of the 17th century, the idea that all organisms exist as result of the replicative processes of like organisms had taken firm hold. The trouble was that it was not certain whether what we now call micro-organisms were organisms at all, or, if they were, what effect their presence had on the media such as air, food, flesh or blood in which they were found.

Even in the early 19th century, most people thought the phenomena of fermentation, putrefaction and disease were caused, not by micro-organisms, but by 'fluids', 'miasmas', 'effluvia' and other indefinable essences which were held to be present in fermentable or putrescible media (see Appendix A).

In 1680 Leeuwenhoek wrote to Robert Hooke: 'In my opinion we can now be assured sufficiently that no animals, however small they may be, take their origin in putrefaction but exclusively in procreation.' (See Plate II.) Leeuwenhoek discussed the generation of mussels from spawn, and the transport of microscopic organisms on air currents (a concept that was to cause a great deal of scepticism among Vitalists until the end of the 19th century). In the same year, Robert Boyle made the semi-observational comment in his *Sceptical chymist*: 'Water produces moss and little worms or other insects, according to the nature of the seeds that are lurking in it.' Boyle, like Joblot shortly afterwards, was persuaded that a knowledge of the causes and nature of fermentation would lead to an understanding of the causes and nature of disease. This idea was pursued by the botanist, Richard Bradley, who referred to micro-organisms as 'animalcules' or 'invisible insects' and wrote (1718): 'Putrefaction, I think, is always attended with insects.' Unfortunately, he failed to convince his contemporaries. This failure can in part be attributed to the progress that was being made during the 18th century in biological systematisation,* and in part to the revival in France of an influential Vitalist movement, which attempted, on the basis of some formidable experimental evidence, to ascribe the putrefaction caused by micro-organisms in unsterilised media to the presence of unspecified vital essences.

* Linnaeus and his pupils placed the agents of disease, not among the insects, as Bradley had done, but with the worms, in a sub-group labelled Chaos, which included the spermatozoa identified by Leeuwenhoek and the vapours of spring clouds (1735). Linnaeus's casual treatment of the disease-causing agents is odd, for he was a Professor of Medicine. But the urge for tidiness is strong and often overrides consideration of adequate scientific knowledge.

TECHNIQUES FOR EXCLUDING VITAL AGENTS

One prominent Vitalist was the eminent French natural histor-
ian, Georges, Comte de Buffon. Buffon's collaboration with
the former Welsh priest, John Turbeville Needham, lasted
from 1748 to 1750 and was attended by considerable publicity.
Buffon was pre-occupied with lecturing, and it appears that
he had poor eyesight; Needham's eyesight was good, but
he had less influence in French society than Buffon had.
It was remarked of the collaboration that, while Buffon
did most of the talking, it was Needham who did the actual
work.

Needham examined under the microscope infusions of
various nutritive substances, such as crushed almonds and
macerated hay, which he had kept for several days in corked
phials. He found that a great deal of activity took place in the
media in his phials, even after they had been heated in hot
ashes. Needham's infusions, teeming with 'animalcules', were
used by Buffon to illustrate his theory of 'indestructible vital
molecules', which were supposedly released on the death of
organisms, only to be reabsorbed later by inert matter to give
life to other animals and vegetables. To Needham and Buffon,
the 'animalcules' present in infusions were preliminary stages
in creation of life from non-living material. We should
probably describe them today as 'sub-vital systems'. The
publicity that attended these suggestions aroused interest in
some scientific circles. In others it aroused suspicion. Voltaire
described Needham as a madman. Needham and Buffon
resolved not to reply to ill-informed criticism, but two disserta-
tions published in 1765 and 1776 by the Abbé Lazzaro Spallanzani, questioning their views, could not be ignored.

Spallanzani found it more satisfactory to view 'animalcules'
as ordinary organisms of very small dimensions. If they were
organisms, their eggs would have to be kept out of putrescible
media, just as Redi had kept flies away from decaying flesh a
century earlier. Spallanzani repeated Needham's experiments
with media such as infusions of beans, beets and egg yolk, but
he hermetically sealed his experimental flasks, which had been

previously sterilised by heating. These experiments were largely successful and formed the basis of the practical activity of food-canning a generation later, for the food in the flasks did not go bad. The Vitalists considered that the rigour of Spallanzani's techniques had rendered the air inelastic and his media too sterile for any type of vital essence to survive. A scientific controversy in which the experimental successes of one viewpoint were the experimental failures of the other was bound to take time to resolve.

To circumvent Needham's objections, Spallanzani drew out the necks of fresh batches of flasks to capillary dimensions. He found in these circumstances that very minute 'animalcules' appeared in previously sterilised media. He suggested that larger micro-organisms had been prevented from entering the flasks owing to the small opening of the capillary tubes. (Later experimenters were to devise various other methods for the filtration of micro-organisms. Pasteur used this same type of flask for some of his own experiments, generally bending the capillary neck to prevent the entry of bacterial contaminants into the experimental flasks against the force of gravity. See Plate III.)

Spallanzani's experiments were not always successful. This was particularly true when he was handling species of micro-organism with a high degree of heat-resistance. Methods of dealing with these were not discovered until the second half of the 19th century. Vitalism survived Spallanzani's attacks, but a number of scientists, such as Joseph Priestley, had been convinced. Priestley (1790) held that it was only lack of knowledge that prevented people from accepting that micro-organisms, like macroscopic forms of life, start from 'organised germs'.

The metaphysical and non-experimental discussion of micro-organisms did not preoccupy practical men, who were quick to appreciate the commercial value of Spallanzani's experimental techniques. In 1810, Appert, variously known as Charles, François or Nicholas, a former distiller, published a paper entitled: *The art of preserving all substances of animal and vegetable origin for a period of years*. Appert's method of preserving fruits, vegetables and meat by putting them, while still hot, in bottles or, later, tins, which were then shut in the manner familiar to housewives, was deservedly popular. The products of this form of sterilisation were palatable and nutritious. It was hard to believe that vital principles had been lost or, if lost, that they were of such fundamental importance as the Vitalists maintained. These techniques soon became the basis of a world-wide industry. Appert's work led to the establishment of a commercial cannery by Donkin and Hall, and to detailed experimentation in many laboratories. It is perhaps as well for the progress of scientific thought that the 20th-century discovery of vitamins lay hidden in the future. Otherwise, the Vitalists might still have believed they were right.

Like Berzelius, Appert thought that he was destroying 'ferments', not killing organisms. He attributed the success of his method to the exclusion of air from the tins and bottles of preserved food, which prevented these 'ferments' from reacting. This idea was pursued in the year of the publication of Appert's paper by the French chemist, Gay-Lussac, who maintained that oxygen (which in Lussac's time had the same general appeal that nucleic acids have today) was the cause of putrefaction. This postulate led the food-canning industry into expensive but fruitless attempts to preserve sterility by removing oxygen from cans of preserved food with vacuum pumps. Lussac's false assumption was identified by Pasteur, who showed that many 'microbes', such as the rod-shaped bacterium which was associated with lactic acid fermentation in milk, were unable to respire at all in the presence of oxygen. This explained why housewives, even though they had no fancy equipment, were making a better job of food-canning than the food technologists.

In 1837 and 1838, Schwann (see Chapter 1), working simultaneously with Schultze, proposed that the putrefaction of flesh was caused by the vital activities of micro-organisms. He bubbled vigorously sterilised air through flasks containing boiled nutrient media, and maintained that sterilisation could prevent the multiplication of micro-organisms. Putrefaction

could, in fact, be avoided.* Together with attempts to eliminate bacterial contaminants went efforts to isolate and culti-
vate micro-organisms. In 1840 von Helmholtz separated a
putrefying medium from an identical but fresh medium by
means of a parchment filter, and showed that the putrefying
agent was retained by the parchment. In 1854 the physiologists,
Schroeder and von Dusch, using apparatus designed by the
chemist, Loewel, to keep air free from 'seed crystals', trapped
bacteria in a long tube packed with gun cotton. Pasteur used
nitrocellulose for this purpose, and then dissolved it in alcohol
and ether. In this way, he was able to examine the micro-
organisms that had been trapped (see Plate III).

Lister pointed out that 'air is filtered of germs by the air pas-
sages'. The turbinate bones in the nose serve the same purpose
of filtration as Pasteur's nitrocellulose. Trapped micro-
organisms could then be inoculated on to previously sterilised
but potentially nutrient media (which were developed with
great success by the German microbiologist, Ludwig Koch) and
studied *in vivo*.

As a result of these refinements in techniques for handling
micro-organisms, more facts became available to the oppo-
nents of the Vitalist view of the causes of putrefaction. In 1859
Félix Archimède Pouchet, Director of the Musée d'Histoire
Naturelle in Rouen, published a book on the spontaneous
generation of micro-organisms. Like Erasmus Darwin half a
century earlier, Pouchet found it incredible that there were
large numbers of micro-organisms floating about in the air, a
belief which had been held by Leeuwenhoek, Spallanzani and
a great number of other scientists. Bent on a spectacular
demonstration of this fact, despite the disapproval of his
former master, Dumas, Pasteur set off to the Alps in 1860. At
the same time, Pouchet took a party up the Pyrénées. Each
party carried with it a set of experimental flasks containing
nutrient media which had been allegedly sterilised before

* Tyndall was to show a generation later that Schulte and Schwann
need not have used such vigorous methods to make the air 'germ-free'.
Air could simply be bubbled slowly through water, which served as an
effective 'germ-trap', or it could be passed vertically up the neck of a
capillary tube, as Spallanzani had proposed.

departure. The flasks were opened on top of a glacier in each case, and then sealed again. Pasteur returned to Paris to report that in the extremely pure air above a glacier there were no bacterial contaminants, so that his flasks had remained sterile. Pouchet reported in Rouen that as many micro-organisms had been generated in his own expedition's flasks as were generated elsewhere. The last encounter between Pasteur and Pouchet was planned for 1864, but Pouchet did not come to the lecture hall.

John Tyndall, who was originally a physicist, was drawn into polemics with the Vitalists as a result of experiments he had been conducting on the paths of beams of light. After 15 years' investigation, he concluded that 'ordinary air is no better than a sort of stir-about of excessively minute solid particles', and he then began to reverse his previous experimental methods and examine the purity of air by the use of beams of light. In England, against the Vitalist, H. Charlton Bastian, Tyndall extended the magnificent polemical campaign that Pasteur was waging in France.

The resistant phase in the bacterial life-cycle was effectively demonstrated in 1876 by the German bacteriologist, F. Cohn, with *Bacillus subtilis* in hay infusions (see Plate IV). Further research into heat resistance in spores was carried out in the 1870s by biologists all over Europe and served as a stimulus to discussions on the origins of life, as we shall see. In 1877 Tyndall read a paper before the Royal Society, entitled *On heat as a germicide when discontinuously applied*, in which he introduced a method of sterilising media containing even highly resistant spores by the intermittent application of heat. Tyndall explained that whereas continuous boiling for one hour might not suffice to make certain species of bacteria inviable, discontinuous heating for five successive periods of one minute each enabled these spores to mature into the heat-sensitive phase of their life-cycles, and so be killed by subsequent heating. Bastian, who believed in 'ultimate living particles', absented himself from the meeting at which Tyndall presented this paper. By 1880, the claims of the Vitalists had been discredited in the eyes of a substantial body of scientists, but

Bastian remained a Vitalist, resistant to the end, until his death in 1915.

THE COSMOZOIC HYPOTHESIS

Once it was widely accepted that germs did not arise spontaneously in decaying organic matter, some scientists began to see possibilities in the idea that life could not ever have arisen spontaneously, but must have arrived on the Earth from another planet. This 'cosmozoic hypothesis' appears to have originated in 1821 with Sales-Guyon de Montlivault, a retired French naval officer, and it has had a surprising vitality. It was taken up with enthusiasm by von Liebig, von Helmholtz, Richter, Lord Kelvin and in this century by Arrhenius. It was the cause of a great deal of discussion and some experiment. The behaviour of spores and seeds was studied under conditions of extreme cold and dryness to see whether they were likely to survive in interstellar space for the time thought necessary if the 'pressure of light' was moving them from planetary system to planetary system. Had scientists been aware then of the intensity of ultra-violet and X-ray radiation in space, the studies would no doubt have been extended to include those agents.

The idea that life had no origin is one of the ways in which what is sometimes called 'the perfect cosmological principle' could operate. According to this principle, the universe has had, apart from local fluctuations, the same appearance at every epoch. If so, there must have been life in it always. This could arise constantly of necessity by one of the processes that we shall discuss, or it could drift around the universe as what Haldane called 'astroplankton', vital units ready to start the cycle of evolution when conditions are favourable. This idea has seemed less implausible since the revival of the theory that the Earth was built up by the accretion of various forms of cold interstellar detritus.

The basic trouble with the cosmozoic hypothesis is that if the universe had a beginning, so, clearly, had life. The idea that life came here from elsewhere skirts around the question of the creation of life and considers only the problem, tricky in its own right, of transferring life from place to place in the universe. Karl Pearson wrote disparagingly in 1892 of 'a meteorite, like an ethereal gondola, which might have brought in a crevice the protoplasmic drop to our Earth. Not much significance need be attributed to this pleasant conceit.'

Soon afterwards Schafer wrote, somewhat acidly: 'The idea merely serves to banish the investigation to some conveniently inaccessible corner of the universe.' It is now generally recognised that, if we are satisfied that life originated anywhere, we might as well start with the assumption that it happened here, where we have experimental material to hand. If we are to be accurate, however, it is clear that we should have to answer the question, 'Where did life originate?', on the lines: 'Here, or anywhere else, provided an environment of suitable constitution is maintained under suitable conditions for a suitable time.' The meanings to be given to the three *suitables* are themes for present and future research.

THE IMPACT OF DARWINISM ON THEORIES OF THE ORIGINS OF LIFE

Late 18th-century scientists, who were in some doubt as to the status they should give to micro-organisms like bacteria and fungi, were in no predicament about the origins of life. Erasmus Darwin thought that moulds were spontaneously generated, and he envisaged a similar generative process in probiotic times as having sparked off the whole process of evolution. He thought life began as some kind of 'primordial filament' and summed up his view in the *Temple of Nature* (1803) with this verse:

> Without parents by spontaneous birth
> Rise the first specks of animated Earth.

The great French biologist, Lamarck, believed in the spontaneous generation of mushrooms, and wrote of 'primeval slime'. He suggested (1820): 'Among the inorganic bodies extremely small half-liquid bodies developed and gradually became organised.'

Such views had become less orthodox by Charles Darwin's

day, for by then it had become apparent to most scientists that the spontaneous appearance of an organism in an initially sterile medium had never been experimentally demonstrated. It was therefore necessary for scientists to postulate an appearance in probiotic times that had not been shown to be possible in the contemporary environment. Pasteur's attitude in this predicament was characteristic. In 1878 he wrote: '*La génération spontanée, je la cherche sans la découvrir depuis vingt ans. Non, je ne la juge pas impossible.*' ('I have looked for spontaneous generation for the last 20 years without success. No, I do not consider it impossible.')

Charles Darwin proposed that structure and capacity had been built up in organisms since life began by means of the gradual and opportunist mechanism of natural selection, and that this process still operated. If he had failed to consider the further question of the origins of life, it would have been tantamount to admitting a major discontinuity in his carefully reasoned argument. He treated this subject cavalierly for the next ten years, and maintained that it was beyond the scope of useful scientific speculation for his day. 'It is mere rubbish thinking of the origins of life,' he wrote in a letter. 'One might as well think of the origin of matter.'

Nevertheless, intense discussion of the question of bio-poesis * began. Scientists like T. H. Huxley, John Tyndall, Allen, Errera, Karl Pearson, Schafer, Herbert Spencer and others wisely refrained from discussing any detailed hypotheses as to the mechanism of biopoesis. This in no way diminished the value of their contributions. They all maintained that organic matter might, under suitable conditions, have been produced from simple carbon compounds and minerals in the

* If we wish to avoid employing ambiguous terms, and to avoid using the expression 'the origin, or origins, of life', we can use the term 'biopoesis'. This describes the creation of something which some people might wish to call 'living' from non-living material. *Poesis* implies 'making', in the sense of creating a new arrangement. Many physiological terms derive from this root; so does the word *poet*, i.e. a 'maker'. *Eobiont* – 'dawn organism' (made by using the prefix 'eo-' as in the geological epoch, Eocene) is a term used to describe historically postulated sub-organisms whose coagulation and evolution is held to have led to the origins of life.

early stages of the Earth's development. They considered that the amount of space and time available in the remote past, together with the different environmental conditions that might have operated, would have made it possible for biopoesis to happen and even happen frequently along these lines, but they thought it was so improbable as to be unlikely in any experimental vessel today.

In 1870 T. H. Huxley addressed the British Association in these terms: 'I should expect [life] to appear under forms of great simplicity endowed, like existing fungi, with the power of determining the formation of new protoplasm from such matters as ammonium carbonates, oxalates and tartrates, alkaline and earthy phosphates and water without the aid of light. That is the expectation to which analogical reasoning leads me, but I beg you once more to recollect that I have no right to call my opinion anything but an act of philosophical faith.' (The chemical basis of respiration without air, known as fermentation, had just been elucidated.) The idea that eobionts might have depended for their source of energy on fermentation was also included in Haldane's theory of the origins of life (1928).

Possibly as a result of the discussion that followed Huxley's address, Charles Darwin left on record his own 'act of philosophical faith', for soon afterwards he wrote (1871): 'It is often said that all the conditions for the first production of a living organism are now present, which could ever have been present. But if (and oh! what a big if!) we could conceive in some warm little pond, with all sorts of ammonia and phosphoric salts, light, heat, electricity, etc. present, that a protein compound was chemically formed ready to undergo still more complex changes, at the present day such matters would be instantly devoured or absorbed, which would not have been the case before living creatures were formed.'

A conceptual framework for a theory of biopoesis was by no means an explanation of a sequence of events. The breakthrough here did not come until the 1920s, though there is no good reason why a theory should not have been systematised earlier. Many of the elements of later theory were recognised. Not surprisingly, as biopoesis was a subject in which he had a particular interest, the works of Bastian expressed views which were later to become orthodox, in spite of the scepticism with which his extensive publications were greeted by prominent Abiogenists of his day. In particular, he suggested that the actinic rays of the Sun might provide a suitable source of energy for biopoesis and that the overlap and coalescence of 'semi-vital particles' might be a necessary step in the creation of life, rather on the lines of the reconstitution of sponges after dispersal, which was extensively studied and discussed at this time.

Table 1. *A summary of the different views on the origins of life by the end of the 19th century.*

Number of biopoeses	General character of biopoesis	Present standing of theory
1. Innumerable	Classical, medieval and Vitalist concept of spontaneous generation	Theory demolished by the development of the food-canning industry
2. None	Cosmozoic Hypothesis. Life has always pervaded space and been transferred from place to place	Awaits testing by space research once diagnostic features of eobionts have been established
3. One	Evolution on Earth by action of inevitable normal processes	Widely held view of many scientists in Darwin's day
4. One	Creation by Divine intervention	Not incompatible with possibilities 2, 3 and 5
5. Several	Repeated co-ordination of sub-vital units, or eobionts	Orthodox theory today

4. 'An Abode Fit for Life'

'Il faut prendre notre globe tel qu'il est et bien observer toutes les parties, et par les inductions conclure du present au passé.' *

GEORGES, COMTE DE BUFFON, 1749

A few discussions can reasonably be expected to end with a definite conclusion. They are organised to consider where or how something happened, or what something was made of, and we can see at the start that there is a general line of research that should produce the answers. Discussions about the origins of life are not like that. Even today, we have little expectation of being able to conclude a discussion on the origins of life with the statement: 'This is how life arose.' The best we can hope for is: 'This is one of the ways in which life could have arisen.' All that we can do is to compare probabilities, and if we are to do that usefully it is essential to keep all the possibilities in mind.

We lack evidence for and definition of the distinction that we should draw between an eobiont and an organism. It is therefore impossible to decide dogmatically how far back in the history of the Earth the origins of life on it should be pushed. We can, however, examine the geochemical evidence for indications of environmental conditions during the lower Pre-Cambrian period. We can also make careful searches among stratified Pre-Cambrian rocks for fossil evidence of the earliest forms of life on Earth. Is it possible that the geochemical and palaeontological lines of evidence will converge sufficiently to enable us to make a clear statement of the problem of the origins of life?

* 'We must take the globe as we find it, observe all parts carefully, and by induction draw conclusions about the past from an examination of the present.'

35

THE EVIDENCE FOR THE COMPOSITION OF THE PROBIOTIC
ENVIRONMENT

In his book, *The age of the Earth as an abode fitted for life*
(1898), Lord Kelvin suggested that the Earth had cooled and
solidified from a hot wisp of solar material more than 20 but
less than 40 million years ago. In the balance, Kelvin placed
his estimate nearer the 20-million mark. The validity of
calculations of this order was doubted by those who considered the thickness of the sedimentary rocks and the rate of
morphological change of which the fossil record gave evidence.
The scepticism was soon justified by the discovery of radio-
activity, which was shown to provide a source of internal heat
for the Earth. (We have to be wary of all quantitative cosmological interpretations until they have been subjected to criticism for a number of years.)

Some scientists still think that the Earth started as a hot
body, but the balance of opinion seems to be that it was built
up by the accretion of small, cold particles about 5000 million
(or 5 G *) years ago. This age has been arrived at by a number
of alternative methods of dating. One method relies on the
assumption that certain minerals present in the oldest rocks
initially contained no lead, but gained it at a known rate from
the radioactive decay of uranium. Another method of dating
relies on the theory that the universe began with all its matter
confined into a compact space, and this matter expanded and
disintegrated, and all its fragments, seen today as nebulae,
have been moving out in all directions ever since: on the
assumption that the velocity of recession has been the same
throughout the history of the universe, we can calculate the
time at which the universe started.

If we believe that the Earth was originally incandescent and
that it had about the same mass as it does today, the oceans
must have formed after it had cooled and the crust became
hard. They must have contained materials derived from those

* G signifies the term 'giga' (10^9), and should be used instead of
'billion', which has a different meaning in the U.S.A. (i.e. 10^9) from what
it has in the U.K. (i.e. 10^{12}).

parts of the Earth's crust that were near enough to the Earth's surface to be washed. (If this theory holds good, the oceans appear to have a surprisingly low salt content, for present-day rivers could bring down that amount of salt in a fraction of the time that the oceans are thought to have existed.) On the accretional theory, a much larger proportion of the whole mass of the Earth was probably under water as a result of the process of condensation. This could indicate that there was initially such a wide range of possible materials present in the waters of the oceans that it would be unwise to be dogmatic about the composition of the pools that first formed on Earth.

On either theory, the atmosphere today must be derived from the original volatile components of the probiotic Earth which had sufficient mass to be retained by the Earth's gravitational pull, together with subsequent exhalations, with which we include oxygen. The oxygen of the atmosphere today is released by green plants as a result of the process of photosynthesis. Each molecule of oxygen tends to be reconverted to water or carbon dioxide, or to combine with the oxidisable constituents of surface rocks within a few thousand years; our present atmosphere would therefore become free from oxygen but for the continuing action of green plants. Herbert Spencer noticed that the total amount of 'fossil' carbon in shale, coal and oil deposits was approximately equivalent to the amount of oxygen in the atmosphere. He then assumed that atmospheric oxygen was originally derived from a burst of photosynthetic activity at the time when these forms of carbon were laid down.

The fact that nothing gains such universal acceptance as a fallacy should make us all wary of unanimity. As the Canadian economist, Vilhjalmur Stefansson, pointed out a generation ago, the picture of the ostrich with its head buried in the sand has taken such a firm hold on our imaginations since Pliny's day, that few of us care to reconsider the responses of the actual bird under stress. There are two reasons for contesting Spencer's view. First, there are large quantities of 'fossil oxygen' sequestered in sedimentary deposits of ferric iron and sulphates. (Plutonic rocks, presumed to be the Earth's oldest rocks, contain unoxidised sulphur and ferrous iron only.) Of the proportion of oxygen now present in the atmosphere, per- haps twice that amount must have passed through the Earth's atmosphere at an early stage in the Earth's history to account for the abundance of oxidised sedimentary deposits such as ferric iron, and sulphates.* Where, if Spencer's view is ac- cepted, has all the carbon or other reducing material gone that would have been simultaneously produced during photo- synthesis?

Secondly, there is a very plausible non-biological source of free oxygen. This is the decomposition of water vapour under the influence of ultra-violet radiation in the upper atmosphere, followed by the loss of hydrogen into space. Lately, the whole question of the origin of the oxygen in the atmosphere has been rendered even more intractable by the suggestion (1961) that the Earth is gaining large quantities of hydrogen each second from the solar wind. If the incoming hydrogen has contributed to hydrogen trapped in the water in the oceans, then some of the oxygen present in sea-water must also be put in the cate- gory of 'fossil' oxygen, along with that in ferric iron deposits, sulphates and other oxidised sediments. If we do indeed exist and have always existed in this hydrogen hurricane, the prob- lem of the origin of oxygen is very much more complicated than we had thought.

If sediments such as limestone and dolomite, which contain the element carbon, were spread evenly over the Earth's surface, there would be about 2000 g/cm^2 carbon 'trapped' in

* Evidence for the differing weathering capacities of the probiotic and biological environment, due to the presence in the latter of abundant supplies of free oxygen, has to be treated, like the rest of the geochemical evidence, with caution. Oxidising and reducing conditions are found today closely juxtaposed in biologically active sedimentary environ- ments such as swamps and areas of marine deposition. Both sulphur and iron are subject to reversible changes in solubility and oxidation states. This would lead us to expect that minerals dating from such an early period as the Pre-Cambrian have very probably been subject to extensive alterations, which would diminish their value as indicators of environ- mental conditions during the original period of deposition. Only a complete lack of iron or sulphur throughout a deposit suggests that reducing conditions prevailed exclusively at the time of deposition.

these minerals. Concentrations of one-quarter to half as much carbon are trapped in bitumen, coal, oil, etc. By no means all this carbon has necessarily been involved in vital processes, but most of it at one time existed as free atmospheric carbon dioxide. The carbon in the sediments alone has carried out of the atmosphere 25 times as much oxygen as is present in the atmosphere today. This, taken in conjunction with the fact that rates of production and sequestration of atmospheric gases are uncertain by factors of at least 10, makes it legitimate to conclude that dogmatism about the composition of the atmosphere at any stage in the Earth's history before perhaps the Jurassic is out of place. All we can say with certainty is that very large amounts of O_2 and CO_2 have passed through it. The concentration of either of these gases at any phase in the early history of the Earth is still a matter of assumption.

There is some dispute, too, about the status of nitrogen, which is today the principal component of the atmosphere. Some geochemists believe that nitrogen was a primary component, others that it was a secondary one derived from the oxidation of ammonia. Nitrogen compounds play a fundamental role in life now, and they may well have been important for biopoesis. But nitrogen gas itself is too inert at ordinary temperatures for it to be important, even if it was a primary constituent of the atmosphere.

The probiotic atmosphere is thought to have been made up of volcanic gases and possibly some residual water, ammonia and hydrocarbons present in the cold interplanetary detritus. Parts of it may have ranged from being alkaline with ammonia to being acid because of the oxidation of sulphur and hydrogen sulphide. Moreover, the effective environment for biopoesis may well have been less extensive than the whole of the probiotic hydrosphere. Indeed, it may have been located in regions that were in some way atypical (see Chapter 7). It is clear that there is at present no basis for dogmatism about the conditions that prevailed in the environment during the period in which we now believe biopoesis was taking place.

SOME PRE-CAMBRIAN FOSSIL EVIDENCE

A number of what appear to be fossil organisms, such as colonial algae and other assemblages of simple organisms with known present-day equivalents, have been dated to a period of 1·5–2 G years, and assigned to the middle Pre-Cambrian period. An example of one of these early fossil assemblages, dating from the lower middle Pre-Cambrian of the Gunflint District, Ontario, Canada, has been described by Tyler and Barghoorn (1965). The principal fossil-bearing beds are some unique, fairly incompressible cherts, in which abundant, unbranched algal filaments, microspores of various shapes and sizes, and one or two other types of organism have been identified. Their morphology is shown in some detail, as they are preserved in optical continuity within the chert, which is believed, on the basis of mineralogical evidence, to have originated as silica gel, accumulating in an iron-rich shallow-water environment subjected to alternating inundation and dessication.

One or two species of the Gunflint fossil micro-organism known as *Kakabekia* have now been shown to have modern relatives with a very similar morphology, consisting of a stipe (stalk) ending in an umbrella-shaped extension. The previously unknown modern relatives of *Kakabekia* have been successfully cultivated in anaerobic conditions in the laboratory, a discovery which might lead the unwary to suppose that here we have direct evidence that the micro-organisms of the Gunflint cherts subsisted at a time when the Earth's atmosphere was an exclusively reducing one, a time in which gases such as ammonia took the place of oxygen in the biosphere.

Any such interpretation is invalidated by the fact that the proportion of C-12 isotope (see Chapter 1) segregated in the algal fragments of the Gunflint cherts is almost identical with the C-12 readings given by much later coal and lignite deposits. This suggests that photosynthesis must already have been established at this early date, a discovery that is of great interest in itself but which offers little comfort to the student of terrestrial biopoesis in search of evidence for the stages interposed between the build-up of complex molecules on an ir-radiated planet and the evolutionary progression of organisms, a period which would appear on present evidence to have lasted at least 3 G years. The fact is that the Gunflint cherts contain an assemblage of organisms with modes of nutrition equivalent to those we should expect to find in any modern micro-environment in which aerobic and anaerobic conditions were closely juxtaposed. (One way round the conceptual problem of establishing a point in time in which we might wish to claim life began, is suggested by the 'double-cone' diagram in Chapter 7 (Fig. 3).)

ORIGINS OF LIFE THEORIES SINCE THE 1920s

T. H. Huxley and his contemporaries assumed that organic matter would gradually be synthesised from the simple carbon compounds and minerals on a sterile Earth. A great advance was made when Haldane (1928) described a plausible mechanism. Herbert Spencer had suggested that most, or even all, atmospheric oxygen had been made by green plants. Haldane pointed out that, if the probiotic world were anaerobic, any complex molecules that were generated would be more stable than they tend to be in the presence of atmospheric oxygen.

Ozone (which is the form in which some oxygen exists in the upper atmosphere and which is a powerful agent in promoting synthesis) absorbs ultra-violet light, shielding the Earth's surface from concentrations of ultra-violet light that would be deleterious to biological systems. Before there was free oxygen in the atmosphere, much more ultra-violet light would have got through, and Haldane deduced that the primitive ocean, in an environment containing neither oxygen nor micro-organisms, would have attained the consistency of a 'hot dilute soup', so that any organism that did manage to appear in it would have abundant food.* He went on to argue that eobionts must have depended, as embryos and many bacteria do, on fermentation. This is a logical necessity for life in an

* Bequerel suggested (1924) that ultra-violet light may have played a part in promoting biopoesis, and located its site in the upper atmosphere.

environment containing little or no oxygen. Haldane's suggestion (1928), that a metastable environment which would have been a necessary prerequisite for the growth and multiplication of the earliest organisms could have been provided by the 'hot dilute soup', gave the question of the origins of life a concrete biochemical form and put a little meat on the skeleton provided by Pasteur, Huxley, Tyndall, Allen, Schafer and Errera.

Another of Haldane's early papers on the theme of biopoesis contained some fruitful speculations about the nature of viruses and genes. He suggested that they might be considered as fragments of organisms which, although unable to reproduce in normal culture media, could reproduce in the specialised environment of a cell. He suggested that the coalescence of somewhat similar 'half-organisms' may have provided a mechanism whereby the prototype of a free-living organism could have been built up. In 1953 Haldane enlarged upon this, and suggested that systems with a quasi-vital degree of complexity appeared independently. This is a reasonable induction and does not require that the specific nature of the original components of the aggregating system be described. (Haldane was one of the first to write of the origins of life in the plural.)

These ideas have retained general acceptance. Haldane often commented that their acceptance in principle in both the U.S.A. and the U.S.S.R. caused him some perturbation because of his mistrust for orthodoxy. Most theorists in this field have also accepted that the presence of an interface of some kind is a necessary prerequisite for biopoesis, be it air/water, mineral crystal/water or liquid/liquid, as Oparin suggested in his study, *The origin of life on Earth*, which appeared in Russian in 1928, and was translated into English in 1938.

The possible role of mineral crystals in biopoesis (see also Chapter 6) was considered in some detail by the English bacteriologist, Twort, working in the 1930s. Twort tried to devise a medium (outside the living cells of the host) for the cultivation of viruses. His attempts failed, but he did appreciate the difficulty of what he was trying to do. He arranged incubators illuminated by light reflected from a galaxy of mineral surfaces, and used a range of special clay extracts to make up his media. Goldschmidt (1952) stressed the catalytic activities that might result from the absorption of components of the probiotic soup on the edges and corners of mineral crystals. Thus hydrated silica, alumina, cyanides, phosphates and carbonates absorb molecules and hold them in a spatial array in a manner comparable to that in which they are held, or assumed to be held, by proteins (see Chapter 7). Goldschmidt pointed out that calcium phosphate and carbonate still play a role that amounts to symbiosis in many organisms. Bernal proposes a similar role for clay. We know too little about the physical chemistry of adsorbed molecules to be able to see clearly which idea is the more advantageous.

The central problem of biopoesis is to discover how a mixture of complex molecules could become sufficiently coordinated to satisfy the minimum requirements for an organism (see end of chapter). In effect, this is the study of processes of evolution in purely chemical systems. Oparin's cardinal contribution to the question of the origins of life has been his analysis of processes analogous to selection and evolution occurring in chemical emulsions. The conditions under which eobionts in the probiotic soup developed and changed can, Oparin suggests, be investigated in the laboratory by the study of liquid concentrates rich in colloids, which he calls 'coacervate droplets'.

These droplets form when a wide variety of substances of large molecular weight are mixed at room temperature. The molecules of such mixtures migrate from the liquid in which they are suspended, leaving it almost free from colloids. The molecules swarm together to form sharply demarcated droplets, or sometimes, as in the case of Leisgang rings, they form layers, differentiated from the surrounding medium. (Such systems have the life-like property of developing an organisation that seems to demand the use of some readily available source of energy; their mode of formation is still inadequately explained.) Oparin suggested that a system of immiscible droplets accumulating on the surface of the primitive Earth could have concentrated various types of complex molecules, which grew and divided until they reached critical sizes.

Under the influence of a form of natural selection ('biochemical evolution') these evolving systems provided the basis for the origins of life. The advantages that flow from juxtaposition gave survival value to more and more complex forms of biochemical organisation, until there evolved systems from which the earliest organisms developed. Synthetic droplets of the kind Oparin describes have not yet been shown to satisfy the minimum vital requirements for an organism. This does not invalidate the theory. They may have been made under the wrong conditions. A search for more relevant conditions would be interesting, and possibly illuminating.

THE DESIGN OF DEVICES TO MEET THE MINIMUM REQUIREMENTS FOR AN ORGANISM

If we can decide arbitrarily on some minimum experimental requirements for a vital system, we may eliminate a great deal of profitless discussion on the definition of life and also establish a yardstick whereby we can distinguish a system that is an analogy for life (see Chapter 2) from one that we will decide to call 'living'.

Primitive forms of life probably depended, as present-day forms do, on a pre-existing energy flux or metastable system in which an energy-yielding reaction can be catalysed. J. W. S. Pringle has devised a chemical model for a vital system based on the oxidation, by oxygen or hydrogen peroxide, of hydrocarbons, produced by the action of water on carbides. If a chain reaction were set up through the participation of free-radicles, such a system would, he points out, have very life-like properties. For example, if some non-vital mechanism maintained mixing and flow of the reagents in a tube, and started combination in it, there would be a zone of combination with properties clearly demarcated from those of the rest of the system. It would be at a higher temperature, would contain substances not found in other parts of the system and would produce a characteristic set of metabolic products. If a

Plate I. '*Moss agate*'. *Dendritic growth form of iron contaminant in chalcedony.* (*Courtesy of Chan Kwok-hoi.*)

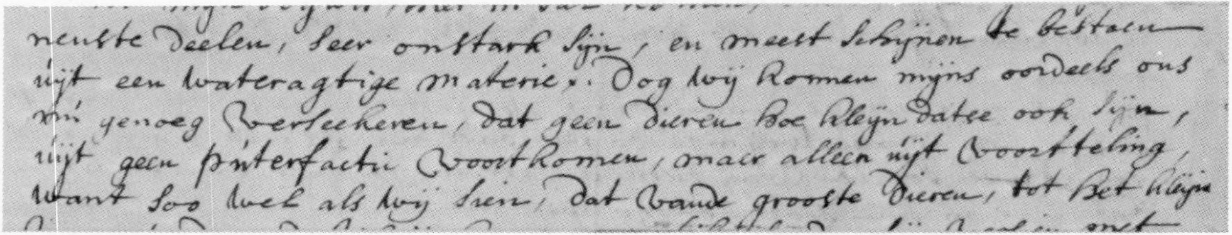
neutte deelen, seer onstark sijn, en meest schijnen te bestaen
uijt een wateragtige materie. Dog wij konnen mijns oordeels ons
nu genoeg verseekeren, dat geen dieren hoe kleijn datse ook sijn,
uijt geen puterfactie voortkomen, maer alleen uijt voortteling,
want soo wel als wij sien, dat vande grootste dieren, tot het kleijn

Plate II. *Facsimile of an extract from Leeuwenhoek's letter to Robert Hooke, dated 12 November 1680.* (. . . ***But, in my opinion, we can now be assured sufficiently that no animals, however small they may be, take their origin in putrefaction, but exclusively in procreation.***) (*Courtesy of N. W. Pirie, F.R.S.*)

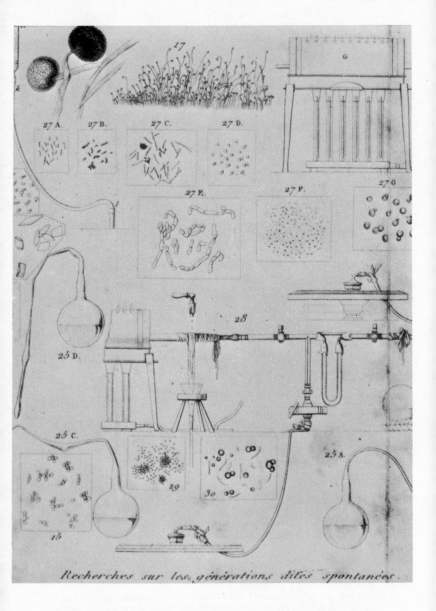

Plate III. *Section of engraving used to illustrate Pasteur's paper of 1860 on the organised particles present in the atmosphere.* (C. r. hebd. Séanc. Acad. Sci. Paris, **348**, 52.)

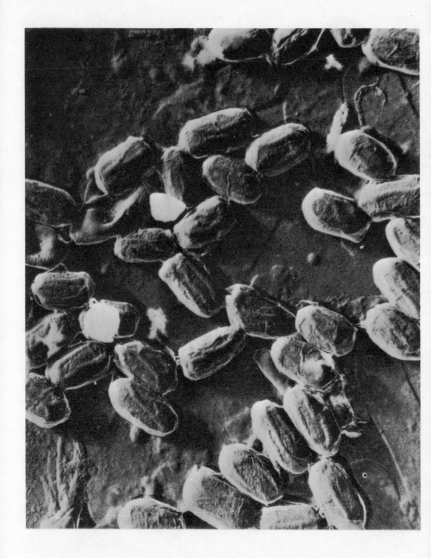

Plate IV. *Electron micrograph of* Bacillus subtilis (× 6000). (*Courtesy of Drs D. E. Bradley and J. G. Franklin.*)
This was one of the bacteria which enabled the proponents of the theory of spontaneous generation to extend the polemics on the question of the origins of present-day micro-organisms from non-living precursors into the second half of the 19th century.

(a)

Plate V. *Our oldest
known ancestors –
some fossils from the
Gunflint (Ont.) cherts
of the lower middle
Pre-Cambrian.
(Photos: courtesy of
Professor Barghoorn.)*

(a) *An algal filament
resembling the pre-
sent-day blue-green
algae. (The two
oblique splits in the
slide are imperfections
in the chalcedony
matrix.)*

(b) *Another algal
filament, accompanied
by smaller
spherical cells,
possibly iron-fixing
bacteria. (× 1130)*

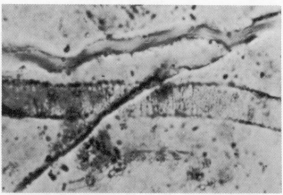

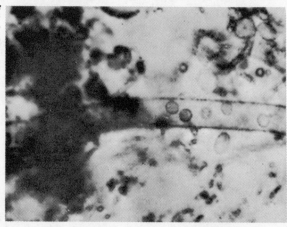

(c)

(b)

(c) *Three views of the micro-organism from the same cherts (× 2000), known as
Kakabekia umbellata. This organism has recently been compared with a contemporary one of similar morphology (consisting of a bulbous stipe ending in a
spoked circular tip) isolated from soil at the foot of the walls of Harlech Castle.
Spores of the contemporary form have been grown on agar in a reducing atmosphere, an environment which effectively imitates the urine-soaked soil from
which it was isolated. (See* Proc. natn. Acad. Sci. U.S.A., 1966, **55**, 349.)

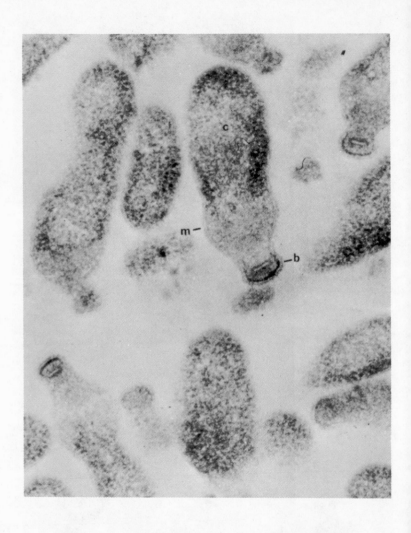

Plate VI. *Our smallest living relative – single cells of* Mycoplasma gallisepticum (× 72 800). (*This magnification should be compared with that of the micrograph of the* Bacillus subtilis *spores and with the blue-green algae fossils from the Gunflint cherts.*) *Key:* m = *double mucopeptide membrane round PPLO;* c = *cytoplasm containing randomly arranged organelles and areas of fibrils in which nucleic acid materials are concentrated, similar to the internal cell constituents of bacteria;* b = *the 'bleb' areas at the end of the elongated PPLO cells, which are highly structured areas containing no nucleic acid.* (*Courtesy of Drs R. J. Barnett, J. Maniloff and H. J. Morowitz.*)

Plate VII. *Alternative chiral arrangements of a single tetrahedral form (top part of diagram), shown approaching a chiral sieve (lower part) to which half are adapted and half are not. (Courtesy of N. W. Pirie, F.R.S.)*

These go through

These do not go through

Plate VIII. (a) *A large right-handed quartz crystal.* (b) *A large left-handed quartz crystal, showing the diagnostic facet (outlined) which may be developed on the vertical prism face to indicate the chirality of a quartz crystal on inspection. (Courtesy of Dr Claringbull and the Trustees of the British Museum (Natural History).)*

transfer of material from the reactive zone were made into another potentially reactive system quickly enough to preserve some free radicles, the new system would become reactive too.

The qualities of a dynamically stable system in which the formation of active intermediates is balanced by their break-down in a continuous and steady flow, go far towards satisfy-ing our criteria for an organism metabolising in a terrestrial environment. This model has the advantage over the old flame analogy in that it operates at temperatures in the present-day biological range. But which out of a large number of pos-sible chemical analogues provided the basis for actual meta-bolic pathways is a less tractable problem. Any picture we may wish to present of the actual steps that led to biopoesis will be more of a parable than a hypothesis.

A suitable mineral surface, kept moist with a solution of the components of the primitive ocean and atmosphere, would absorb light and promote a reaction whose product dissolves the active element, such as copper or iron, from the mineral. If, by water retention, the substance attached to the crystal promoted that leaching, we have at least an analogue for an organism. This process is in a strict sense autocatalytic.*

Autocatalysis would be favoured by agents that improve illumination or keep the product in the neighbourhood of the mineral. The accidental simultaneous presence of another catalytic system using the same element and making a sponge or an oil would be absorptive or dispel dust and so permit better access of light. Such a conjunction would spread, and pieces of it scattered on to suitable surfaces would start new focuses. Even two actions linked thus show the beginnings of organisation. More could be added if they occurred in the neighbourhood and favoured those already present. Each action proceeds independently, but each is favoured by others.

To put our minimal experimental requirements for a vital system in the most general terms, we would all probably agree that it should be or should contain liquid, and should work at a temperature below, say, 200° C; gaseous systems and systems

* An autocatalyst catalyses the reaction that produces it (see Chapter 2).

working at red heat carry us too far afield. It could contain any type of molecule, but, while we would accept most readily systems containing water, protein (see Chapter 7) and the conventional organic molecules, there are extra difficulties in interpreting results with such systems. The easiest course open to a sceptic will be to explain away all claims for biopoesis as the result of contamination by existing organisms and, as effort is expended on the search for newly formed vital systems, it is likely that a rich harvest of filter-passing phases of bacteria and of heat-resistant spores will be reaped. The more conventional the medium, the more likely contamination becomes, as was emphasised by the numerous attempts to refute the Vitalists' claims for spontaneous generation.

The system must be able to do something. Simple growth is not enough; crystals do that. Nor should activity start as soon as the components are mixed, if the system is to conform to most rigid standards. We should expect to have to wait either for the occurrence of the random process that is being investigated or for the 'seeding' of the mixture by another example of the system: that is to say, we should insist on the ability to reproduce. (The ability of the enzyme trypsin to convert the protein trypsinogen into more trypsin was mentioned in Chapter 2.)

No known example of an organism, regardless of its size, exists by virtue of a single catalytic process. We do not know what the minimum number of enzymes is (see Chapter 5). But while we do not need to demand the ability to catalyse hundreds of actions, we might reasonably insist on five or six. (Haldane suggested four (1957).) Some of these actions should be concerned with the synthesis of the catalysts themselves. If the replicating system by which more trypsin is produced were able to make trypsinogen as well, when supplied with amino acids and sugar or with amino acid derivatives, it would be difficult qualitatively to defend its exclusion from among the systems which satisfy our minimum requirements for an organism.

Many will not choose to look upon a system of this kind as part of biology at all, but it is not until we consider biological systems in such generalised terms that we become aware of the sort of problems we are likely to run into when we consider biopoesis. The fact remains that in spite of the plethora of information that has accumulated, we have as yet no more basis for confidence on the subject of biopoesis than had Darwin, Tyndall or Huxley. That the original culture medium in which life developed contained organic matter is certain. That it would not support life as we know it today is probable. Meanwhile, the factors that lead to different conclusions about the probiotic atmosphere lead to different conclusions about the surface.

Further research will set a limit to the range of possible compositions for the Earth's surface. We might therefore argue, with Darwin, that it is still premature to attempt to map out any routes from a little known geochemical past to an as yet unelucidated biochemical present. There are two important reasons for disagreeing with such a postponement. The fact that scientists are working on the problem of the origins of life is an important stimulus for the accumulation of evidence about the probiotic environment, which may have implications both for industry and for space research. Reciprocally, if biochemistry can unequivocally define the necessary qualities and composition of an eobiont, this would, at least in part, shed some light on the probiotic environment, rather as the morphology of fossils is used to define whether the region in which the original organism lived was dry land, marsh or water.

5. Antecedents

'When we look at minute living animals and see their legs, which are ten thousand times thinner than the hair of my beard, we are bound to assume that these legs, besides bearing instructions for movement, are also provided with vessels for carrying food.'

ANTON VAN LEEUWENHOEK, 1670

Leeuwenhoek thought that the internal structure of micro-organisms represented, in miniature, the structure of more familiar organisms. Today we recognise the uniformities of biology in terms which have little in common with those of the 17th century, but the reason for these uniformities in organisms from widely scattered parts of the biological hierarchy remains one of the central questions of biology. Why is it, for example, that there should be such a striking similarity between the membranes and organelles of so many living cells? Why do all the organisms so far studied use similar optically active molecules? (See Chapter 6.)

The uniformities of biology led late 18th- and 19th-century evolutionists to believe that all organisms were descended from a single ancestor. Ethnologists involved in the study of human artefacts were inclined to be more open-minded. They recognised that while the use of the same decorative symbol by two communities could be taken as evidence for culture contact, the use of the wheel could not, because there was no alternative that was equally effective. An example of a compound of biological importance which is probably used out of necessity is the carotenoid pigment present in plants and animals. Plants use carotenoids to 'trap' quanta of light energy. These are then used for the synthesis of carbohydrates. Animals use the same chemicals to activate the nerve cells that set in motion the processes that enable them to see.

48

It is clear that we would be wise, when considering the problem of biological uniformity, to bear in mind the factor of common origin. We should also take into account the part played by structural or functional necessity in the evolution of biological systems.

In this chapter, we shall be considering the reason for the existence of an apparently limiting size for all organisms able to survive in the present terrestrial environment in the absence of other complex molecules.

THE MINIMUM COMPONENTS OF AN ORGANISM

Leeuwenhoek compared the blood corpuscles he saw under magnification to sand grains scattered on black taffeta. This led his biographer, Dobell, to conclude (1932) that he must have used some form of dark-ground illumination, and that, if this were so, with a magnification of about 200–300 diameters, he would be able to distinguish objects larger than one micron (μ). This magnification would have allowed him to distinguish most forms of bacteria.* The smallest bacterial cocci known are about half this size, and have volumes of about $0{\cdot}1\ \mu^3$. Free-living organisms smaller than this are extremely rare and very demanding in their nutritional requirements, as we shall see later in the chapter.

Soon after the recognition of the viruses, when very little was known about the structure or dimensions of the colloidal molecules that constitute living cells, Mackendrick (1901) and Errera (1903) calculated, on the basis of the concentration of the elements nitrogen and sulphur in the cells of the smallest bacteria, that a micro-organism must contain at least 1000 protein molecules if it was to be able to carry out normal vital activities. This estimate is now known to be inadequate. First, it underestimates the number of complex molecules a cell of

* The only group of micro-organisms Leeuwenhoek missed was the viruses; but this is hardly surprising (as we saw in Chapter 1), because the criterion still generally used to decide whether an infective agent is a virus or not is that it should be invisible even with a modern microscope. In any case, we are not including the viruses among the organisms, for to do so would be to rob an otherwise useful category, 'micro-organisms', of its homogeneity.

volume 0·1 μ³ could contain. Recent calculations (Pirie, 1963) have shown that a free-living organism near the limiting volume, made up of about 8% dry matter, could contain at least 10 000 enzymes of average weight 1 Md,* and many enzymes† are less than one-tenth of this weight. Ten thousand enzymes would seem more than adequate to account for vital activity. It is clear that we shall have to look for some other restriction on size.

Progress in cytological research has shown that one of the fundamental restrictions on the size of micro-organisms must be the relationship which exists between the enzymes and the integrative mechanisms of the cell. The most relevant integrative structures are the organelles known as ribosomes, or microsomes, which are scattered in the cytoplasm of even the smallest micro-organisms. (We are unable to discuss organisational units like the nuclei and mitochondria of more complex cells in this context, as they are large or larger than many of the smallest organisms we are considering.)

The ribosomes are a heterogenous group of particles. They are isolated or clumped together in the cytoplasm and associated with several different complex molecules, including ribonucleic acid. Their diameters are about 15–30 nm, and it has been calculated that there is room on each one of them for about 14 protein molecules of average weight 35 000 d. At least 20 or 30 of them are believed to be crucial to the organisation of the metabolism of most, if not all, micro-organisms

* Md = one million daltons (the dalton is one-twelfth the mass of the most abundant carbon atom).

† Enzymes, without exception, have molecular weights in the thousands. There is a dramatic increase in catalytic efficiency in a series of enzyme models of increasing size. The enzyme itself is more efficient than any of the models. There is no satisfactory explanation for this, nor for the fact that all enzymes so far purified use the architecture of a protein to attain this apparently necessary size. It is possible that size is important because, simply as a result of inertia, it stills Brownian movement. It is also possible that size is important because it allows one molecule partly to encompass another so that contact is sufficiently prolonged for chemical changes to take place in the absence of the close bonding that is necessary when small molecules interact. It is clear, then, that there is already, at the enzymic level, a limit set to the minimum size of a cell.

able to survive on their own in the predator-filled contemporary environment.

In the presence of suitable substrates and energy sources, the ribosomes seem to be responsible for the synthesis of different proteins such as enzymes, under the influence of different instruction filaments. Protein synthesis is at present pictured as taking place on each unit of a row of microsomes through which a ribonucleic acid filament is drawn. This filament is thought to control the sequence in which amino acids are added to growing protein molecules in much the same way as the holes in a pianola roll control the tune.

We shall now consider what restriction the enclosure of the cell contents in a relatively impermeable cell membrane must impose upon the possible size of micro-organisms. (Some form of enclosure was considered an essential feature of a minimum organism by H. E. Armstrong in 1912.) A cell membrane helps to preserve the ordered arrangement of catalysts. It enables them to act in sequence, bringing about progressive changes in their substrates, or adding successive pieces during the synthesis of complex molecules. The membrane protects the mechanism of the cell from the external environment into which metabolic by-products are excreted and from which essential substrates are withdrawn. (J. B. S. Haldane, who preferred to keep an open mind about the complexity of minimal system, pointed out that a containing membrane was not an essential requirement of 'laboratory life'.)

What is the thickness of the thinnest biological membrane that has physical coherence? There is an interesting uniformity about biological membranes, regardless of their origin. Thus the thickness of the outer membranes of all unicellular groups of organisms is similar to that of the membranes surrounding most mitochondria, nuclei, and forming the Golgi lamellae in some tissues from the vertebrates. (Rabbit mitochondria seem to have slightly thinner membranes.) All these measurements are subject to the uncertainties of electron micrography, because the sections measured may not always have been cut at right angles to the surface of the structure being sectioned. Moreover, staining, which is necessary for the effective examination of tissues, may distort the thickness of the membrane. Certain types of staining exaggerate the width of an object, while others tend to diminish it.

The thinnest biological membrane that has physical coherence, however, seems to have a thickness of at least 5 nm. This uniformity is impressive and so is the three-layered appearance of the membrane; this is generally interpreted as two protein layers separated by a layer of lipid. Mucopeptides are responsible for the rigidity of bacterial cell walls. If the mucopeptide structure is ruptured chemically by a suitable enzyme, the cell contents can be seen ballooning out through the ruptured segment of the cell membrane. Cytological studies indicate that a mucopeptide membrane needs various types of reinforcement if it is to be effective, for it is built up of an open-net structure, with holes in the net through which particles as large as 0·1 Md can pass. This diameter is greater than that of a co-enzyme, and as the biochemist, M. Dixon, remarked (1949), the loss of co-enzymes 'is what is really meant by the death of the tissue'. *

We can conclude that in order to retain essential co-enzymes, the synthesis of which may be carried out on parts of at least 20–30 microsomes, an organism must have a relatively impermeable cell membrane at least 5 nm thick. These minimum structural requirements would appear to be reasonably fulfilled by micro-organisms with a volume of at least 0·1 μ^3. Simple organisms of these dimensions would depend on complex organic molecules such as amino acids and other metabolites from the external environment, together with externally supplemented energy sources. It may be that the rarity of smaller micro-organisms is simply a reflection of the rarity of unoccupied environment niches supplying these things.†

Haldane considered the minimum size of organisms in the

* The leakage of essential catalysts into the medium surrounding the cell has already been shown to destroy metabolic activity. A similar type of leakage has been demonstrated with the viruses. Thus the Rickettsia viruses lose ribonucleic acid and co-enzymes. With this loss they lose their infectivity. When the co-enzymes are restored, the viruses recover infectivity.

† Microbiologists estimate that 20 times the present biomass exists in the form of bacteria as exists in the form of animals.

light of quantum physics. He reminded delegates at a con-
ference (1963) on the origins of life: 'We are not bold enough
in our physics. We are thinking in terms of the physics of 1920
rather than the physics of 1963.' Noting that physicists had
already established a principle of increasing order in matter
from electrons to atoms, and from nuclei and electrons to
molecules, Haldane suggested: 'There are far more complicated things of the same general kind in connection with living
systems that help them preserve their structure and function.'

Simplified arguments based on quantum theory are somewhat nebulous, as our habits of thought are attuned to matter
in bulk and not to atoms and quanta, but Haldane argued that
the application he proposed was no more metaphysical than
the conventional explanation of the phase change from gas to
liquid. In gases, the molecules interact very little. In liquids,
they interact thoroughly. Quantal events are spread through
space and time without parts. They are extended in time in
inverse proportion to their energy. The less the energy, the
greater the extension in space, as we know from the behaviour
of electromagnetic radiation. Haldane was interested in the
way in which separate quantal events will overlap. It was this
overlap, he suggested, that enabled the phenomenon we term
life to emerge. With rapid events taking place in an atom,
overlap is rare. There would be significant overlap, however,
with the slower and less energetic events in a cell, which have a
considerable extension in time, permitting a higher degree of
interaction than on an atomic or simple molecular level.

THE SEARCH FOR THE SMALLEST FREE-LIVING ORGANISMS

We can now consider the group of Mycoplasmataceae, some
members of which have volumes which fall below the minimum volume for a free-living organism (0.1 μ^3). (Many
PPLO are elongated, and fall within the size range of the
smallest bacteria. This is true of the cells in the electron
micrograph shown on Plate VI. The ribosomes are scattered
in the cytoplasm of the cell, and the double walls of the cell
membrane can also be identified.) PPLO, such as the one

responsible for Pleuropneumonia, were isolated because they caused disease in man and domestic animals. Subsequently, less pathogenic **PPLO** were isolated because a deliberate search was made for organisms smaller than bacteria in the rich, nutritive medium of sewage. How certain can we be that we have not missed organisms even smaller because they do not obtrude themselves as pathogens or because they fail to show some active form of metabolism?

The viruses, though not strictly relevant to this discussion, serve to illustrate the point about difficulty of discovery. It would be difficult to recognise the presence of a virus that did not cause disease. It would be carried in the host without symptoms and would have no metabolic activity. Only a very painstaking search for differences in the macromolecular constituents of many different putative hosts of the same species would reveal the 'organism'. Similarly, the search for small organisms in an environment such as the sea or soil would involve looking for some change initiated in inert media by inoculation. This would be feasible by electron micrography or by studying changes in chemical composition, but the search would be exceedingly tedious. If there are organisms with virus-like dimensions in the sea or soil they would quickly, in this size range, become attached to one or other of the ample absorbent surfaces that are suspended in it, such as the surfaces of mineral particles, detritus or other organisms.

Whatever future searchers may find, the PPLO are the smallest micro-organisms so far cultivated *in vitro*. These minute organisms are unusually exacting in their nutritional needs; they grow slowly and attain only a small population density at the end. There is disagreement whether these phenomena are a consequence of inadequacies in the medium, the presence of toxic components or products of PPLO metabolism, or of an intrinsic enzymic inadequacy. It is also possible that the poor growth of PPLO agents can be ascribed to the fact that, though they may be able to make co-enzymes, they cannot survive unless they are able to retain them, or be supplied with them at suitable concentrations from highly complex external media such as blood.

THE RELEVANCE OF THE STUDY OF THE SMALLEST ORGANISMS TO
THE PROBLEM OF THE ORIGINS OF LIFE

Those apparently simple bacteria that flourish in media containing simple molecules (known as 'autotrophic bacteria') are of unknown antiquity. Most scientists believe, however, that they do too much to be relevant to the study of eobionts. They have a biochemical expertise that is greater than that of man or any other animal, and this is only likely to have been achieved by a process of gradual evolution. Viruses, organelles, genes and various other small constituents of present-day cells do too little to be relevant to the study of biopoesis. Their capacities, like those of the autotrophic bacteria, are likely to have been slowly accreted in response to millions of years of competitive stress. If we grant that these small systems are sophisticates rather than primitives, it is clear that there is no reason to assume, as several scientists have in the past, that eobionts were of necessity small. We may, indeed, be wiser to counteract this simplifying assumption by picturing eobionts not as minute particles at all, but as systems extending over the surface of a mineral crystal, or even over an area as large as an acre and through many intermediates.

Since the simplest organisms, like the PPLO, have the smallest range of metabolic activity known short of dependence on the actual constituents of host cells, they serve as useful pointers to the nature of life. Their dependence on complex substrates is not in itself any disqualification for their relevance to the study of eobionts. Most scientists agree that there were plenty of complex molecules at hand in the probiotic environment. The difficulty is that the most satisfactory interpretation that can be placed on the origin of the viruses and of the PPLO agents is an interpretation based on the secondary devolution of these systems from less dependent organisms, organisms capable of existence in the absence of the highly sophisticated substrates on which viruses and PPLO depend.

If we compare the original systems from which the viruses and PPLO may have devolved, with the 'beast machine' of

Descartes, we can see that there are two opposite ways in which their complexity could have been diminished:

1. The mechanism of self-construction could have been preserved, while the ability to make the components could have been sacrificed. This would lead to the type of machine which can assemble itself from a limited number of pre-existing parts. Such a machine would be analogous to an organism of the PPLO group in that it contains the integrative mechanisms necessary for vital processes (e.g. ribosomes), but is unusually exacting in its nutritional requirements.

2. Simplification could proceed, on the other hand, by the elimination of the integrative mechanisms. This would leave only a set of instructions that must be followed by other similar machines which will make, not new machines, but new copies of the instructions. This line produces an analogy with the viruses. Both modes of approach start with a fully independent system and whittle it away. It seems reasonable to suppose that PPLO and viruses originated in this way. It follows that no similar process could have operated during biopoeses, because the probiotic Earth could not, by definition, have contained systems ripe for whittling away.

If this analysis is correct, the study of the smallest organ-

Table 2. *Approximate* sizes of small organisms and their minimum components.*

Unit membrane	7·5 nm
Ribosomes	8–20 nm
Plant and animal viruses	20–300 nm
Smallest known PPLO (possibly unable to replicate)	125 nm
Probable functional size for PPLO	1μ
Small bacteria	$1\mu \times 0{\cdot}5\mu \times 0{\cdot}5\mu$
Usual bacteria	$1\mu{-}2\mu$
(These bacterial widths are general, though the lengths are variable.)	

* These measurements have to be taken (in ascending order) by X-ray measurement on dry crystals, by electron microscope after drying and shadowing or by light microscopy after fixing and staining. All these methods subject the 'wet' form in which these systems normally exist, to considerable interference. The measurements given can therefore be taken as an approximate guide rather than as a fixed scale.

isms, however interesting from the practical standpoint of pathology or the theoretical standpoint of the essential nature of life, is a dubious guide to the conditions operating during the extended period of evolution during which vital systems began to emerge.

6. The One-Handedness of Nature: The One-Handedness of Life

'No one walks on alien soil.'

PLOTINUS, c. A.D. 260

In the first century B.C., Lucretius considered the relationship between the two possible arrangements that can be made when a fourth object is added to a group of three. The fourth object, he saw, can be added either above or below the plane defined by the other three objects, thus forming a tetrahedral structure. By interpreting the combining powers of carbon (and other 'quadrivalent' atoms) in terms of the anchorage they could provide for four suitable chemical groups, the German chemist, Kekulé, in 1858, was proposing a tetrahedral structure of the kind Lucretius had considered.

The idea of a three-dimensional chemistry was revived by the chemists, Le Bel and Van't Hoff, independently in 1874. They postulated that an organic molecule like that of the D-tartrate shown in Fig. 2 must of necessity exist in two forms, each of which is the mirror image of the other. The relationship that exists between two structures which are mirror images of each other will be termed 'chirality'.* The term was proposed by Lord Kelvin in 1884, and has been used until recently in the field of experimental physics. The word is

---

* The concept of chirality, or 'one-handedness' is similar to that of 'isomerism', which was proposed by Berzelius in 1830. Isomeric substances have different properties even though they are made of the same elements in the same proportions, because the elements from which they are constructed are differently arranged. The term 'optical isomerism' indicates the relationship between isomerism and the practical way in which it is identified in crystals and solutions by the manifestation of optical activity (see Chapter 1).

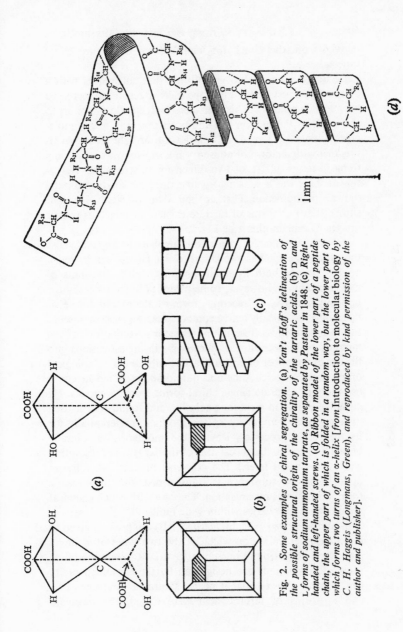

Fig. 2. *Some examples of chiral segregation. (a) Van't Hoff's delineation of the possible structural origin of the chirality of the tartaric acids. (b) D and L forms of sodium ammonium tartrate, as separated by Pasteur in 1848. (c) Right-handed and left-handed screws. (d) Ribbon model of the lower part of a peptide chain, the upper part of which is folded in a random way, but the lower part of which forms two turns of an α-helix [from* Introduction to molecular biology *by C. H. Haggis (Longmans, Green), and reproduced by kind permission of the author and publisher].*

1nm

derived from the Greek for 'hand', since hands are chiral objects.

Le Bel and Van't Hoff established that substances with a chiral bias must inevitably appear as a result of increased chemical complexity. This idea gave mathematical support to the work of Pasteur on the ability of optically active molecules to rotate the plane of polarisation of light. However, it was met with a storm of abuse from chemists who could only think in terms of flat, or two-dimensional, structures. Nevertheless, here was a satisfactory structural explanation for the optical phenomenon Pasteur had demonstrated. He had shown that two forms of tartrate existed, arranged as mirror images of one another (see Fig. 2).

All ordinary chemical operations involve so many molecules that, if a molecule has two chiral forms, chance can be relied on to produce both in equal amounts within the limits of measurement. If, however, some means is found whereby the optically inactive, or 'racemic', form of a compound is separated into its optically active components, the process of separation is known as 'resolution'. Pasteur's 'resolution' of the tartrates by inspection was followed by his investigations into the use of a number of optically active substances of biological origin, such as cinchonine and quinidine; he found he could use these to separate single chiral forms from racemates. He kept live mould in a racemic solution and found that it used up one chiral component of the racemate preferentially for its metabolic processes, leaving the mother-liquor chirally biased. (Chemical agents capable of similar types of resolution will be discussed later.) The converse process, whereby an optically active substance is transformed into the racemic form, is known as racemisation. The ease with which chemical systems racemise varies within wide limits.

D- and L-(*dextro* = right, *laevo* = left) are prefixes given to molecules, depending on which of the two types of arrangement they have around one chiral centre. Racemic modifications of compounds are designated by the combined prefix DL-. Pasteur showed that if nearly saturated solutions of D- and L- forms of tartaric acid were mixed, the racemate crystallised out and the temperature of the solution rose. This would suggest that the opposite process, resolution, might be thermodynamically impossible, because the free energy of the solution of a racemate would appear to be smaller than that of either of the isomers. There is, however, no contradiction; the energy of crystallisation is available to drive the reaction in one direction. This is a source of energy that is reasonable in theory, and it has already been invoked to supply larger amounts of energy in practice.

It may also seem improbable that the macroscopic manifestations of chiral imbalance represented by a large left-handed quartz crystal, or by solutions of cane sugar or egg white (which were found to be optically active soon after the techniques for polarising light were established at the beginning of the 19th century), should succeed in becoming massed together to the exclusion of the opposite type. Chiral molecules, however, solve this problem for themselves. They often do it when they crystallise. Each crystal is built up by the nearly perfect packing together of molecules of one or the other type, or of both types in equal numbers. There are no haphazard assortments. This again follows of necessity if space is to be filled economically.

When we pack chiral objects together in a box we generally have to avoid haphazard arrangements for the same reason, if we are to do it neatly. Hands illustrate the point well. For certain gestures, the Indian gesture of greeting, for example, a right hand touches a left. The gesture would be clumsy if it were attempted with two right or left hands. A handshake illustrates another form of packing. It is only possible with a pair of right or left hands. The difference depends on the point that Lucretius understood; in one arrangement the wrists point in the same direction from the chiral centre and in the other they point in opposite directions. As we shall consider, the same necessity for perfect or almost perfect packing determines the way in which even very complex associations of molecules, known as polymers, are linked together.

In the limit, initial selection of molecules with a single chiral bias can be carried out by a very simple structure, which could be called a chiral sieve.* To take a crude macroscopic analogy (see Plate VII), this sieve does not need to be any more complicated than a metal plate punched with regularly orientated slots. Chiral sieves organised to accept a single chiral arrangement of molecules find it difficult or impossible to accept the opposite form. Suppose we picture a molecule approaching a chiral sieve as if it were a dandelion 'parachute' (or seed) falling through the air under the influence of gravity. The structure will be kept orientated because it is feathery at one end and weighted at the other. At a certain level of complexity, only a suitably shaped object will pass through a particular orientation of slot (Fig. 2). The orientation of the object will be maintained by gravitation and air friction, or some similar constraints. Molecules approaching a chiral sieve are subject to various other constraints; a charged surface orientates any ion approaching it, for example. The orientation of the slot will be a predetermined structural characteristic of any organism, and will be one of the factors determining the possible responses of a cell to an incoming substance. An incorrect orientation of object will be rejected immediately, though a chiral sieve will have a somewhat greater toleration for variation in size.

The reason why organisms use only one of two possible chiral alternatives is comparable to the reasons why engineers use right-handed screws almost exclusively: it is tidier. If both chiral types were used indiscriminately, a much more extensive range of tools and parts would have to be kept in the workshop and many more trials would be needed to find the right one. In a competitive world, a workshop that standardises itself on one type of screw is likely to prevail; and organisms live in a competitive world. Thus it is easy to understand why an organism tends to use one set of isomers exclusively.

* The fundamental mechanical processes involved in the selection of single isomers are well illustrated by the Penrose 'parabiont', described in Chapter 2. These models show on a very simple level that once an accidental arrangement has been achieved, the system acts to perpetuate the arrangement.

DIRECTING AGENCIES IN NATURE

A superabundance of biased non-vital agencies could have been present in the sterile environment in which life originated.*

Pasteur suggested that the hydrocarbons made when steel dissolves in acid might be optically active if the steel were magnetised. Others have suggested that the Earth's magnetic field might provide a source of asymmetry. In some of his experiments, Pasteur tried to make plants synthesise unusual isomers by illuminating them with light from a heliostat driven so as to reverse the normal movement of sunlight. Like a number of scientists since, he called attention to the rotation of the Earth and to the consequent regular movement of the direction from which light comes. This effect, it has been claimed, might produce a slight excess of right-handed polarised light from such sources as scattered, circularly polarised moonlight or ultra-violet light. The search for potential polarising agents of non-vital origin has continued, in spite of the lack of experimental success which has characterised much of the work.

There are also numerous optically orientated mineral interfaces present in the surface rocks (see Chapter 4). If any chemical action depends on a catalyst that is chiral, it is reasonable to expect the catalyst to introduce a bias. The natural enzymes present in organisms are biased and act in this way. Model, or artificial, catalysts made in the U.S.S.R. with D- or L- quartz crystals as base, coated with catalytically active materials, have also been shown to produce a chiral bias in various solutions.

Many people have singled out quartz as the actual orientating agent during biopoesis, for the curious reason that it is very abundant. It makes up about 15% of the lithosphere, so

* There is, of course, no reason to assume that synthesis in the absence of a physical or chemical directing agency cannot give rise to a set of molecules with a preponderance of one chiral form. As the number of molecules that is being synthesised gets smaller, the probability of an imbalance between the two chiral types gets larger, until in the limit only one molecule is being made. Of necessity it can be of only one type. This is obvious, but it seems to have perplexed many people.

that if a pool of a racemate should be formed it would be very likely to get seeded quickly. To be effective as an organiser of the spatial separation of the two isomers, however, a seed must be extremely rare, so that the probability of the pool being seeded by both forms will be remote. It is clear, then, that it would be unwise to postulate that the natural abundance of any directing agency, be it quartz, clay or right-handed circularly polarised light, increases the likelihood of this particular agency having been the only cause of the present chiral preferences of materials of biological origin.

Goldschmidt has pointed out, indeed, that quartz might be brought into the argument for precisely the opposite reason. An experienced instrument-maker told him (1952) that large right-hand natural quartz crystals were commoner than equally large left-hand crystals (although small crystals of either type were found equally often). Goldschmidt suggested that this might be because optically-orientated organic matter, synthesised after one type of isomer had become dominant on Earth, might have suppressed the growth of left-hand crystals. The essence of Goldschmidt's suggestion is that probiotic molecules and their descendants may have modified the environment enough to influence the chiral bias of certain minerals that have crystallised there since.

By 1900 it was clear that there are so many polarising agencies in nature that, given a bias, however slight, it is reasonable to argue that competing eobionts or organisms might swing to one extreme or another. The existence of even a slight trace of asymmetry was all that was required to initiate an extensive chiral bias in probiotic times. Probably the most widespread method for the effective spatial separation of isomers in nature would have been the random scattering of crystals and mother-liquor during or after crystallisation, which could give spots with a bias in favour of one or the other isomer. Once crystallisation of one or the other of two isomers began, the process of crystallisation might have been interrupted and the mother-liquor drained away, leaving a local concentration of one chiral form. Similarly, if both chiral forms crystallised separately and the crystals were subsequently scattered, each crystal could serve as focus for the segregation of a single chiral form. In the 1890s, unnecessarily dramatic agencies such as volcanoes were sometimes invoked to bring about the separation of probiotic isomers, but the possibility of isomers being separated was clearly recognised; this possibility is, nevertheless, periodically rediscovered and published by people with an inadequate knowledge of the literature of their subject.

By this period, scientists who did not accept that the building up of a preponderance of one type of molecule is no more intrinsically improbable than the initial synthesis of a complex molecule, were beginning to interpret the origin of the marked chirality of biological systems in terms of structural chemistry. These investigations produced some significant results. Crystallising systems were known to organic chemists by the end of the 19th century in which the preponderance of a single chiral form of an easily racemised substance evidently increased at the expense of the mother-liquor, provided a suitable stabilising agent was present. Pope and Peachey (1900) found that derivatives of methyl-ethyl-propyl tin racemised so fast in solution that when one isomer crystallised out, the other did not accumulate, but the whole preparation instead finally appeared in one form.

The fact that an atom of tin could act as the centre of a chiral structure attracted so much attention that the significance of racemisation as a possible means of getting bulk quantities of one isomer tended to be overlooked. Because D-camphor-sulphonic acid was the agent used to stabilise the D-isomer from the crystallate, many tended to look on this as an example of the normal separation of isomers by combination with another optically active substance. In principle, any other substance that made a readily crystallised complex with the tin compound should have had the same result. This aspect of the work was not pursued, and it is only during the last 14 years that the gap is beginning to be filled.

Tri-O-thymotide racemises rapidly in solution, but it forms adducts with non-chiral molecules such as benzene, and these are stable. When crystallisation starts at one focus, the crystal is either D- or L- and the whole preparation separates in whichever form started the crystallisation. The excess of the other form remains only briefly in the solution and racemises in the same way as the tin compound. We have therefore a mechanism whereby unequivocal *dextro-* or *laevo-* rotation can be built up in racemising solutions. Obviously, either isomer is equally likely to be the one to start the process. Unconnected centres will probably be different. Therefore, in the presence of suitable stabilising agents (which need not themselves be chirally biased), a whole environment can be converted to a single chiral type without the other accumulating at all.

Polymer synthesis has made significant contributions to our ideas about the part that intermediates may play in the segregation of single chiral forms from rapidly racemising systems. Polymerisation is the process whereby two or more molecules of a substance unite to give a larger molecule which has a similar composition but a greater molecular weight. It is now customary, as we have seen, to assume that polymer synthesis may have been the basis of the formation of the aggregates of complex molecules that provided the starting-point for biopoesis.

Meagre as knowledge is in this field, there are already some relevant facts. Organic chemists studying the synthesis of amino acid polymers, such as the helically-twisted polymer of poly-*y*-benzyl glutamate from its corresponding anhydride, have shown that the helical configuration is built up most readily when a growing helix is supplied with carboxy-anhydride components that are either exclusively L- or D- in form. To use a crude analogy, the process of polymerisation can be compared to the zipping action of a zip fastener. The mechanism is only effective when the metal teeth along which the zip will be fastened are regularly aligned in the appropriate direction. The rate of formation of poly-*y*-benzyl glutamate from the corresponding carboxy-anhydride is maximum when only the D- or L-glutamic isomer is present. In the presence of 5% of the alternative isomer, the rate of formation falls to $\frac{1}{3}$, and in the 50:50 mixture to $\frac{1}{17}$ of the rate that is possible in the presence of a single isomer. Clearly, therefore, the 'wrong' isomer is not merely ineffective for chain building (for we would expect that to diminish the rate by only 5 and 50% respectively), but is actively obstructive.

If the formation of any sub-vital system depended on the existence of a polymer of this kind, it would clearly be favoured by any environment that had contained a single isomer. Furthermore, although mixed polymers containing both D- and L-glutamic acid are built up, they are less stable than pure polymers, in particular they are less well able to maintain the well-packed helix that is characteristic of so many biological structures (see Fig. 2). If this is a general phenomenon among polymers, it looks as if organisms are only perpetuating a structural necessity that preceded them.

CHIRALITY AT THE ATOMIC LEVEL

Pasteur maintained throughout the polemical battles with the Vitalists that the division between vital and non-vital systems was clearly a field for further investigation. His researches into the induction of chirality by external directing agencies had been unsuccessful. He subsequently felt obliged to interpret the causes of biological disymmetry in terms of more fundamental forces; in 1874 he wrote: 'What is the nature of the disymmetry of vital actions? I personally believe they are of a cosmic order.'

This statement seemed nebulous at the time it was made. Some experiments, widely publicised in 1960 and still under investigation, show that although we should expect the properties of fundamental particles to have no bias, the emission of certain products from low-energy nuclear reactions consistently indicates that there may be a built-in 'screw' in atoms themselves. This would suggest that the preferred directions of chiral imbalance in biological materials today could conceivably be a consequence of inherent nuclear preferences in the atom.

Haldane suggested that the only way this thesis could be disproved was if the D- isomer of an amino acid (or any other biased molecule opposite in form to those found in terrestrial biological systems) was found in a sample of biological material returned, say, from Mars. (The L- amino acids are used predominantly by terrestrial plants and animals, though there are exceptions, as we shall see.) If L- amino acids were found on Mars, we could draw two inferences: either there had been a coincidence during Martial and terrestrial biopoesis or our life and that of Mars must have emanated from the same source. The discovery of an alien chiral form of a familiar biological compound would, however, invalidate the Pasteur–Haldane hypothesis.

The chiral bias of molecules of terrestrial biological origin is marked in the proteins. No protein is known with the D-isomer of an amino acid in it. (It may be that so many of the early claims for their presence have been dismissed that too little effort is now put into the search.) The exclusion of D-amino acids is not, however, complete; they occur in peptides, alkaloids and other products of many bacteria and fungi – penicillin, for example. Among other examples of chiral versatility, the use of both isomers of lactic acid and some sugars may be mentioned. It is organisms farther down the evolutionary scale which are the most versatile and tend to use D- amino acids. This suggests, though perhaps not very strongly, that selective pressure has not yet been able to force these organisms into the economical states attained by those that are more highly developed. The chiral uniformity of the plant kingdom is better established than that of microorganisms, and is sufficiently complete in the flowering plants for them to be generally useful as forage for protein building in animals. (When they are not useful, this is because they contain substances that are toxic.)

Reciprocity is part of the nature of life. All the vital systems with which we are familiar depend to varying degrees on other vital systems, with which they co-exist or which preceded them. Any bias appearing in an eobiont, whether it rose accidentally or as a result of natural polarising agencies, would have been quickly extended. (Pasteur and Haldane put forward the suggestion that chirality may even be characteristic at a subatomic level.) Any organism that appears in an already biased environment stands a far better chance of survival if it conforms with the chiral preferences which surround it.

7. 'A Wider View of Life'

'In attempting to pursue the evolution of living matter to its beginnings in terrestrial history, we can only expect to be confronted by a blank wall of nescience.'

E. A. SCHAFER, 1911

The early days of biochemistry were marked by an unwilling-ness to countenance more complexity than the chemical evidence demanded. The pioneering biochemists thought mainly in terms of small molecules with determinable structures. Substances such as vitamins and glutathione, which have relatively simple structures, were in conformity with this view. By the 1930s there was a certain amount of disquiet as to its validity. Osmotic pressure and ultra-centrifugal measurements made it increasingly necessary to think of proteins in terms of 200 or more amino acids; Emil Fischer had thought that a figure of about 20 would be sufficient.

It was in this atmosphere that research on the nucleic acids began. Four bases were found in these acids. It was assumed that each of the four bases appeared only once in each nucleic acid molecule. At the time it would have been gratuitous to assume more. In this way the 'tetranucleotide hypothesis' was born. The hypothesis was supported by the assumption that a tetranucleotide was about as big as a molecule ought to get. This assumption sprang in turn from the novelty of the idea of the determinable size of molecules during the early part of the century. The tetranucleotide hypothesis had the serious practical effect of delaying the acceptance of the specific role that nucleic acids must play. Even ten years ago there were those who argued that nucleic acids could not be important carriers of genetic or other forms of specificity because they could not be complex enough. The number of known components in nucleic acids has shown a welcome tendency to increase in recent years. More than a dozen bases are now known for them.

The tetranucleotide hypothesis was succeeded by another simplifying assumption, that the high degree of specificity characteristic of vital processes resides, and has always resided, exclusively in the canonical 25 amino acids, and five purines and pyrimidines. The truth probably lies, as it so often does, somewhere between the exclusive biochemistry envisaged recently and a more catholic range of molecules anticipated by the earlier biochemists. Indeed, when we are considering the origins of life, it is to this latter range of chemicals that we may primarily have to turn our attention.

Invocation of a single solution to any complex problem may well blind us to possibilities we would be wiser not to ignore. On the basis of the widespread assumption that life can only exist as a result of the activities of proteins, nucleic acids, high energy phosphates, or whatever type of compound is at the moment fashionable, a great deal of discussion on the origins of life has centred on the structure, metabolic peculiarities and enzymic activities of proteins and nucleic acids.

Many scientists seem to get a great deal of emotional satisfaction out of neat little designs in which deoxyribonucleic acid (DNA), ribonucleic acid (RNA) and protein are arranged in patterns connected to each other by suitable arrows. It is probable that this is a useful intellectual stimulus, but these designs are simply up-to-date equivalents of the Gnostic symbols; their bearing on contemporary problems is similar to the bearing of

```
S A T O R
A R E P O
T E N E T
O P E R A
R O T A S
```

on the problems of an earlier age. We cannot assume that any single substance is the *sine qua non* of life. Nor, however much it might seem to clarify our problem, have we any good reason

to assume that the initial appearance of any particular substance in its present-day form was the signal for life to begin.
Although writing now generally involves paper, we would not
assume that paper was necessarily its original vehicle. Detailed
discussions of the behaviour of complex molecules of biochemical significance (as with the case of the characteristics of
the PPLO agents) may have no more to do with the origins of
life than a study of the mechanism of a cigarette-lighter has to
do with the origins of fire-making.

One of the greatest advantages the proteins offer as a basis
for living lies in their colloidal properties, and in the possibilities they offer for specificity. Their merit as vehicles for
vital activities may lie in the housekeeping economy they make
possible, for they are built up in much the same way as an
alphabetical language, which produces an almost infinite
variety by ringing the changes on 26 letters.

There are innumerable non-protein catalysts which have
actions similar to those of proteins. Oxidations can be catalysed by many metals and by thio-urea. Some of the rare-
earth elements are esterases. Hydrated silica, alumina,
cyanides, phosphates and carbonates can absorb molecules
and hold them in a spatial array in a manner comparable to
that in which they are held, or assumed to be held, by proteins. It is fair to add that no one has yet constructed even a
model organism out of colloids such as silicates, polyphos-
phates, cyanides, or chelating agents such as ethylene diamide
tetra-acetate and the metaphosphates, or oxygen carriers such
as rubrene. But neither have we yet managed to construct a
convincing para-organism out of the chemicals which compose
most present-day organisms, such as proteins, fats, polysaccharides, nucleic acids and so on. A para-organism constructed out of biologically atypical catalytic systems might be
sluggish, but it would be conceivable.

The vertebrates use proteins for nearly all purposes, structural as well as catalytic. As we descend the evolutionary
scale, we find that the proteins become far less dominant,
their structural and protective functions being taken over
increasingly by polysaccharides and molecular frameworks

loaded with mineral crystals. The proteins do not offer possibilities of reactivity, specificity or structural coherence that could not have been got earlier in, for example, the polysaccharides. The organisms we are familiar with tend not to use polysaccharides in active roles to anything like the extent that they use proteins. But this is a fact that requires interpretation; it is not logically derivable from anything that we know otherwise about these two groups of substances.

The bugbear of discussions on the theme of the origins of life is the attempt to draw up prematurely detailed schemes in terms of present-day ignorance. If we are to make any useful contribution to the problem, all we can do is to conclude: 'This is one of the ways in which life could have arisen.' All we can do is to compare probabilities, while keeping in mind as many possibilities as we can.

Haldane introduced some clarity of thinking into the subject. He gave reasons for believing that it would be possible to calculate the probability of the appearance of a degree of complexity and activity that could justify the use of the term 'living'. In his first article on this theme (1952), he estimated that an ocean containing a variety of organic molecules would by pure chance probably meet the specifications calling for between 50 and 200 bits* in 1 G years. He estimated that a bacteriophage contained about 100 bits. In his last article on this theme, Haldane revised this assumption, suggesting that even one enzyme of the present-day type would need 100 bits, a suggestion which implies that the origins of life may have taken place over an even longer period of time, for Haldane estimated on this basis that 5000 bits would have had to evolve before an artificial 'organism' could have appeared. The disparity between these two estimates depends on the capacities we feel we should demand of an organism (a problem which was discussed in Chapters 4 and 5).

   * A 'bit' is the unit of information or of choice between two equally probable alternatives. Thus the selection of one object (or course of action) from among four requires two bits, from among eight, four bits and so on. It can also be regarded as a unit of negative entropy.

THE APPEARANCE OF SPECIFIC COMPLEX MOLECULES UNDER SIMULATED PROBIOTIC CONDITIONS

Biochemists, like other people, find what they are looking for. They have accumulated laboratory evidence that amino acids, sugars, purines and pyrimidines, the units from which the nucleic acids are made, together with many other complex organic molecules of biological importance today, can be synthesised under random conditions from such simple mixtures as water, methane, ammonia and hydrogen, or activated hydrogen cyanide. The fact that products which are the vehicles of life on Earth today can be so formed is very interesting. Its significance is commonly misinterpreted, however.

One conclusion we can draw from these experiments is that most of the complex molecules found in organisms could have been present in the probiotic environment. However, it is important to combat two common assumptions about such experiments. They cannot be held to demonstrate either that the primitive atmosphere was a gas mixture similar to that used in any particular laboratory synthesis or that the compounds synthesised in these experiments were of necessity components of all eobionts. These laboratory syntheses provide us with random samples of possible probiotic syntheses, nothing more.

For the past 40 or so years many biochemists have been prepared to assume that if a theory of biopoesis demands the initial presence of any particular type of compound, there would be no justification on present knowledge for denying the possibility of its probiotic synthesis. Further research will produce a vast range of types of molecules that might be formed from a different range of substances in different possible environments, and that could then interact. From among these reactions it will be possible to map routes along which analogues for life could have proceeded. This route mapping will be slow and intricate, and by the time we are well embarked on it there may seem to have been some loss in the spontaneity of spontaneous generation.

So far, probiotic syntheses have justified the biochemists'

expectations. An important reason for continuing work in this field is that, with modern methods of separating complex mixtures, these experiments may provide a cheap source of useful organic compounds for industry.

THE SEARCH FOR EVIDENCE FOR BIOCHEMICAL EVOLUTION

Evolution operates to give us morphological structures that are relatively simple and efficient without by any means being the only possible structures; it would not seem unreasonable to conclude that a similar process of simplification operated on biochemical structures to produce the efficient enzymes with which we are familiar.

Biochemical capacity today can be looked on as the successful end result of what would appear to be a phase of biochemical evolution which preceded the Pre-Cambrian and lasted for a period that, on present reckoning, would appear to be about as long as the period of morphological evolution that followed it – that is to say, over 2 G years. By the Cambrian, organisms were probably at the top of the biochemical tree. Since then a few have made jumps or flights, but most have been able to do nothing but climb part of the way down again, losing, rather than gaining, biochemical capacity. During most of the period, the most difficult problem for an organism has not been the maintenance of structure and activity at the expense of food and energy sources in the environment, but the avoidance of being used as a food itself by voracious neighbours. As Darwin and others pointed out in the 19th century, before life became distributed so fully over the Earth's surface this was not so, and it is reasonable to assume that at this time there were more possible modes of existence open to an organism.

Morphological elaboration permits biochemical inadequacy. It may even encourage it. An organism at the primitive stage, at which it is a simple bag that contains enzymes, depends on metabolites diffusing into it and must make everything it needs from whatever materials may come to hand. This puts a premium on biochemical efficiency and adaptability. An organism of this type would need the un-specialised genius of a successful Robinson Crusoe.

Along with morphological and mechanical evolution goes a greater independence from the environment, because the organism can increasingly recognise favourable or noxious conditions and arrange to enjoy or avoid them. For a mechanically highly-evolved organism, biochemical expertness loses some of its survival value. We see this in the greater synthetic capacity of rooted or non-motile plants, as compared to the more generally motile animals. With the development of intellect, our mechanical and physical adequacies also lose some of their survival value.

There is only one apparent site of increase in biochemical capacity at present, and this is in protein synthesis, as Haldane pointed out in 1954. Proteins seem to be able to take on more elaborate forms in more advanced organisms. If this inter-pretation is correct, it would be compatible with a gradual rather than an initial assumption of importance by the pro-teins. If we assume that every difference could be biologically significant, the potentiality the proteins offer for specificity is vastly in excess of anything that organisms can use. A peptide chain that contains 120 amino acids of 20 types could have 20^{120} arrangements, and the possibilities of configura-tional difference could be increased still more by folding and cross-connection. An organism weighing 70 kg could contain only about 10^{23} of such protein molecules. Most of the two million species that now exist are smaller than this, and it is unlikely that they have been preceded by 10^{100} more primitive species. There is, therefore, a fantastic superfluity of specificity available. Every protein molecule that has ever existed on Earth could have been different. It is hardly surprising to find that even in their composition, to say nothing of their con-figurations, proteins of biological origin are less variable than they might be.

Other definable biochemical changes are simplifications, and suggest we are at present tending towards a minimal state of affairs rather than an ultimate one. This trend towards simplifi-cation is apparent at both ends of the biochemical scale, both with those complex types of molecule now most intimately associated with living, like the sugars and purines, and with the so-called bio-elements. Marcel Florkin has shown, with various complex molecules, how often a substance that seems perfectly adapted for one role appears in more primitive organisms in an entirely different one. Thus the hormone oxytocin exerts uterine control in mammals, but controls water metabolism in amphibia. The more primitive animal species have the most complex fats. This may be a consequence of their failure to fractionate food fats, but that in itself is evidence of versatility. Some insects use substances, such as formic acid and hydrogen peroxide, that seem to fall outside the range of normal biochemistry. They may well have been doing this for 100 million years. The simpler organisms, such as moulds and bacteria, handle a more diverse group of metabolites (the use of the D- isomers of amino acids was mentioned in Chapter 6) and produce a more diverse group of excretory products than any other group of organisms. (Unfortunately, the evidence for the antiquity of these groups is uncertain.)

There is no clear evidence, as Schafer recognised, as to which compounds and elements were significant for the metabolic processes of eobionts. The uniformity of biochemistry is frequently emphasised. But if we wish to present anything approximating to a complete balance-sheet for organisms to-day, we would be wise to investigate the anomalies, which exist even at the species level. About one-third of the naturally occurring elements have now been included among the so-called bio-elements with which present-day organisms unenterprisingly carry out 99·9% of their activities. Many of the contemporary organisms which use atypical bio-elements seem, however, to come from orders of respectable antiquity, as we shall consider. This could indicate that the use to which the bio-elements have been put has changed with time.

One field of research which might shed some light on this matter would be the intensification of geochemical studies, with a view to the detection of non-random segregations of elements now associated with life in ancient rocks. These elements might have stayed at the site of deposition. The search for such segregations could include elements commonly associated with life today, such as calcium, carbon, copper, iron, manganese, phosphorus, vanadium, zinc and so on, and should be extended to include elements such as germanium, molybdenum, nickel and titanium that may well have been vital vehicles at one time.

Oil and bitumen regularly contain vanadium.* This suggests that it may once have been used commonly, as it still is by the group of marine invertebrates known as the tunicates. A primitive species of tunicates are known that have blood corpuscles containing large amounts of vanadium and also normal or still stronger sulphuric acid. An existence dependent on a strongly acid solution containing an element as rare as vanadium is obviously hazardous and precarious. More highly evolved species, able to do the job with a more accessible element such as iron, have an advantage. Dependence on the use of vanadium in quantity seems to have been disappearing. If the vanadium-using tunicates had become extinct, the presence of this metabolic peculiarity would not have been inferred from the fossil record. We seem to have here the end of a biochemical story that has lasted several hundred million years. (Selection does not operate in the same way against the use of traces of vanadium. In traces, the element is present in many organisms. It is essential for some moulds and algae, and probably for other plants as well. At the species level, it has been found that some individuals of the tunicate, *Molgula manhattensis*, replace the normal vanadium among their trace elements by the use of niobium.)

Primitive plants seem to use aluminium and silicon more extensively than do the more highly evolved ones. Aluminium is essential for the growth of some plants and is widely distributed, particularly in the more primitive plant species such as tea, and Lycopodium. Selenium is widely distributed in rocks,

* Secondary specific absorption confuses the interpretation of all evidence of this type (as we considered in the case of the optical activity of petroleum samples in Chapter 1). It is, nevertheless, important to bear in mind the possible significance of this sort of evidence, to ensure that some effort is put into finding and co-ordinating the facts.

so that dependence on it would rarely be a serious disadvantage. Yet it is poisonous to most species. However, a few plants grow better in the presence of selenium. In these plants, the selenium is built into their protein in place of sulphur, so that the protein is poisonous to animals. The phenomenon, if not the interpretation, was known to Marco Polo, who observed that certain species of plants were toxic to mules.

Halogen metabolism offers some other hints about the early course of evolution. Halide ions have probably been available ever since liquid water formed. There may have been free chlorine in the atmosphere. Chlorinated hydrocarbons are present in bitumens both of terrestrial and meteoric origin. The ability to handle the halogen-to-carbon bond may therefore have been useful to an eobiont. Halogen metabolism in corals, sponges and moulds is known to be more enterprising than it is in higher animals. Sponges still make extensive use of bromine compounds. So do the various species of mollusc that make Tyrian Purple. At least one group of plants, *Dichapetalum*, makes fluor-acetic acid. Several of the moulds make a range of chlorine compounds. The vertebrates have almost given up this kind of metabolic expertise, but not entirely. We still use iodine in the thyroid and depend for normal life on this otherwise exceptional synthesis.

There is no reason to think that this use of iodine is new. It is found in present-day amphibia and fish, and may well be a relic of an initially more catholic approach to metabolism. Evolution may not have eliminated it because there are so few regions where the amount of iodine in rock is so small as to make this dependence a disadvantage. The only logical alternative is to look on the vertebrate's thyroid as a relic of that kind of co-existence characteristic of organisms which are described as 'commensal'. The thyroid, on this basis, may be the remains of a commensal sponge that lodged in the gill arches of a primitive vertebrate, in a relationship as intimate and essential as that which characterises the interwoven algae and fungi which form the gross morphology of the lichens.

Life considered throughout its history would appear to be

the result of two quite independent processes; the random appearance of capacities in small atypical regions, and the integration of these capacities into mechanisms adapted to survival in the environment that exists more generally. The former may have been concerned predominantly with reactions

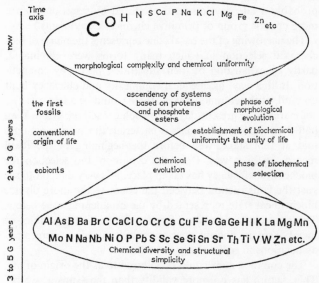

Fig. 3. *At each level of time the width of the cone represents the number of ways in which living or life-like systems worked. The size of the atomic symbol is an indication of the contribution that element may have made to the process at the time. The latest additions to this list are caesium, lanthanum, lead, niobium, scandium, tin, tungsten.* (*Reproduced by kind permission of N.W. Pirie, F.R.S.*)

that are now unusual, but they are the processes that should be taken as the origins of life. They were probably local phenomena.

This view of the evolution of life can be represented by a pair of cones placed apex to apex (Fig. 3). The vertical axis represents time. The base is the group of eobionts, dependent on a very wide range of different chemical actions and energy sources. These modified the original environment on the Earth and, in the course of time competed for sites, energy sources, substrates and so on. This period was characterised by many separate biopoeses leading to families of independent eobionts using different raw materials and reactions. (The use of proteins, built economically out of about 25 amino acids, was probably a late development, produced by selection from a more chaotic group of primitive chemical experiments.)

The narrowing of the basal cone represents the phase of biochemical selection in which many forms were eliminated, partly by exhaustion of their substrates, partly by competition. In this way, biochemical complexity and efficiency built up until, at the common apices, there would be a few varieties comparable to present-day autotrophic bacteria. More than half the present-day biomass is bacterial: this picture assumes that at that stage, well before the beginning of the fossil record, all of it was. Once the evolution and selection for biochemical efficiency has taken place, it is very difficult to see past the construction between the cones to the more diverse biochemical state represented by the broadening cone below. The only clues may come from some surviving oddities of metabolism in ancient species, and from some elementary associations in sedimentary deposits.

The common apex could be referred to as the origin of life. This picture has no more validity than the Gnostic symbol disparaged earlier in the chapter; equally it has no less validity. From the apex of biochemical capacity the upper cone broadens, and we have the evolution of morphological complexity. We have evidence of this phase from the fossil record, and have recognised for some time that it was accompanied in general by the loss of biochemical capacity.

A NOTE ON EXOBIOLOGY

Other environments: different possible ways of being alive

No one has so far demonstrated extraterrestrial vital systems to the satisfaction of any reasonably sceptical person, even though an extensive literature has accumulated on the subject.

While there is no direct experimental basis for the study of possible extraterrestrial systems with life-like attributes, astronauts may soon be able to bring back information as to the suitability of alien environments for life.

If life-like systems do exist in alien environments (we could name such systems 'xenobionts'), it is of immediate and mundane importance to ensure that returning astronauts do not bring contaminants back with them. If there are xenobionts with a form of metabolism completely different from our own, they will probably be poorly adapted to this environment and not therefore troublesome. If, however, the course of evolution has followed slightly different, though comparable, courses in different environments, an introduction of xenobionts could be catastrophic, for the end products of evolution could be similar enough to be pathogen on one another.

The idea that a different group of chemicals from the ones which assumed prominence on Earth could be the vehicles of xenobiology can serve as a valuable corrective to any assumptions we might hold about the nature of life. Some astronomers postulate an atmosphere of liquid ammonia and methane for Jupiter, which has a surface temperature of $-140°$ C. Processes comparable to some form of gaseous life (see Chapter 2) could take place in the very dense atmosphere surrounding this giant planet. In such an environment, complex molecular catalysts like the terrestrial proteins would be less useful than they are here. Here they are useful, for they shelter reacting metabolites from Brownian movement and increase the chances of uninterrupted reaction. In the low temperatures that exist on Jupiter, Brownian movement would not be jogging the catalyst's elbow all the time. The reactions on which 'life', if any, would depend, would in this case be slow. The universe is not short of time, so that this is not a fundamental objection to the idea of life on Jupiter.

Thoughtful Jovial scientists, if such there be, have probably concluded that nothing resembling life can exist on Earth except possibly in the middle of Antarctica. On Earth, the sort of molecules that evolution may have selected on Jupiter would be too stable for life. The basic illusion is to think that ours is the only way of living. On a hotter planet than ours, molecules in our view too stable for life might have been used, though if the reactions ran much faster than ours, specificity would be upset by random fluctuations and movement.

We may also be making an invalid assumption if we think of the processes involved in biopoesis, even on Earth, in terms of contemporary chemistry exclusively, as Haldane argued (1954). He calculated that, in Pre-Cambrian times, the rate at which energy could be liberated by a system resembling present-day muscle would have been so slow that a land animal would scarcely have been able to crawl. The idea that our form of life could arise only when the universe had evolved a chemistry to suit it should be borne in mind, but any judgement on the question may for the moment be suspended.

The possibility of apprehending the communication systems of extraterrestrial intelligence has interested speculative scientists, though attempts to detect such systems have so far not met with success. Haldane examined this idea in the light of quantum physics in three papers he published between 1956 and 1964. The suggestion most relevant to this discussion is that 'thought' may exist in the dense matter of a white dwarf star. It may be impossible to visualise thought without a thinker to communicate with, but if any such system should be recognised, we could not begrudge it the title 'living' merely because it is constructed from unconventional material.

Is there life on Mars?

While the opportunities for speculation about exobiology are considerable, actual facts are hard to come by. Among the most tantalising instances of this are the interpretations that scientists should place on the apparently seasonal waves of colour-change that pass towards the Martial poles annually. The Russian biologist, Tikhov, ascribed these to seasonal plant growth, and accumulated a certain amount of spectrographic evidence in favour of this view (1909). His view was contested by the physicist, Arrhenius, who preferred to ascribe the shifting bands of colour to non-vital causes. He suggested

(1912) that the darker areas were caused by hydroscopic salts. More recently, changes in the state of hydration of some iron compound have been suggested.

Many scientists today favour Tikhov's view. He expected the Martial landscape to resemble the Siberian tundra. (Water seems to be very scarce on the surface of Mars.*) The hope is that we may soon be able to land a television camera on Mars. Such a system is already in use on the Moon and has suggested that the lunar surface is barren (though this does not mean that life-like systems may not exist in the lunar dust-cover, as Oparin pointed out recently).

Contemporary views about the composition of the Martial atmosphere suggest that photosynthesis, if it exists there, does not proceed along conventional lines. But it remains the most probable prime-mover in any biological system. On Mars, the process of photosynthesis may be reversed, so that oxidised material collects in the 'plant' and reduced material is discharged. If such a system exists, it will need surfaces to absorb light and to collect and discharge metabolites. There is no reason to assume that the surfaces will be arranged on small structures. The surface of Mars has probably been suited to the existence of something analogous to life for longer than the surface of Earth, and there is no evidence that the trend of evolution here has been making plants smaller. Therefore, if organisms exist on Mars, there is no reason to expect them to be either small or primitive. Something that looked like a bush would give more convincing evidence for life on Mars than any amount of biochemical analysis.

If water on Mars is scarce, a sparse arrangement similar to a terrestial desert is probable. If the landing site for the television transmitter can be chosen, there would still be a good chance of seeing something. If anything with the mobility of an animal

* The dense, white clouds of Venus, however, have been shown to contain about 1·6% water vapour, and 90–95% carbon dioxide. Their nitrogen content is not yet determined, but is certainly below 7%. A trace of oxygen (0·4%) has also been detected. Part of the interest of these figures lies in the light (if any) they shed on the nature of the original culture medium in which life developed on Earth.

exists, it is likely to investigate the instrument and thus increase the probability of being recorded.

So much for systems that may satisfy our gross aesthetic requirements for organisms, and may be accepted as life-like because they do life-like things. Simpler systems will be more difficult to categorise because, as we saw in Chapter 1, there are no agreed criteria for an organism. Growth, even cyclical growth and shrinkage on a surface, will not be sufficient: something like a lichen would not be visually distinguishable from a hydration or corrosion. If life is not organised in bulk, it probably cannot be recognised without sampling. Having got a sample, we then have to use aesthetic or morphological criteria.

The recognition of organic matter in a sample of surface material from Mars will tell us no more than that there are carbon compounds on Mars. It could even be argued, by analogy with certain terrestrial environments, that the more organic matter that a sample of the Martial surface contained, the less likely is the presence of life, because organisms are likely to be as potent an agency elsewhere as here in tending to restore organic matter to the inorganic state.

Are there any particular complex chemicals we might look for in uncontaminated Martial samples that would indicate the possible presence of some form of life there? A search for such complex chemicals as the high-energy phosphates would have to include the search for various types of esters. It is now an orthodox part of biochemical speculation about eobionts, to maintain that metaphosphates may originally have filled the role of adenosine triphosphate in bringing about phosphorylations on surfaces of a different nature to the enzymes now used. A few years ago, the recognition of protein-like substances in a Martial sample would have been accepted as adequate evidence that there was life there. The abundant chemical evidence that has accumulated since the early 1950s on the formation of amino acids and protenoids in simulated probiotic conditions suggests that the formation of protein-like substances is not always the result of vital processes.

The recognition of optical activity in a Martial sample

would, however, be more significant, as significant perhaps as the recognition of certain types of clearly evident micro-morphology. There are sufficient potential polarising influences to account in a non-vital way for the appearance of a chiral imbalance near the limits of detection, as we have seen in the previous chapter. It would be difficult, however, to account for the appearance of a nearly pure isomer, a terpene, for example, without invoking life. (Terpenes are volatile plant products found in hazes on Earth. If they are released by 'plants' on Mars, even in smaller amounts than they are on Earth, they might serve as an ultra-violet filter, protecting life on Mars rather as the ozone layer protects terrestrial life.)

Haldane did not, it appears, express firm opinions in writing on the validity of the suggestion that there is life on Mars, but in 1929 he did say that Phobos and Deimos seemed better evidence for it than the so-called canals. The density of these satellites is so low, and they are so near the parent body, that it seemed to him possible that they had been made by intelligent beings of some kind. This suggestion was also made by Manhattan (1953), and is one of a very large number of suggestions about the possibility of alternate forms of life in different environments about which we are obliged to keep an open mind.

The universe contains a vast range of process and types of compounds. It is probable that none of these appear exclusively in living systems, but it is equally probable that many more have appeared or still appear in them than is generally assumed. We only know our own kind of life today. We can only describe how to recognise it in imprecise terms, applicable in some instances and not in others. The possibility that eobionts or alternative forms of life may exist or may have existed in extraterrestrial environments, or may have existed, dependent on some form of biochemistry radically different to our own, in terrestrial environments prior to the beginning of the middle Pre-Cambrian, should be borne in mind. It serves as an invaluable corrective to any assumptions we might care to make should we attempt to draw up any rigid operational definition of that very general term, 'living'.

Appendix A

Homer had recognised the need for protecting corpses from flies, as we see from Achilles's speech after the death of Patroclus: 'I would avenge Patroclus immediately were it not that flies might settle on his wounds and cause decay' (*Iliad*, XIX). Techniques for drying, salting, pickling, spicing and embalming were known from the earliest times, as was the fact that putrefaction could be postponed by the use of pies and pasties with a continuous, fairly dry casing, surrounding a moist heat-sterilised core. Contagion was clearly recognised in the Bible, as we see from the precautions advised in Leviticus against contact with lepers. Fracostoro, in the 16th century, compared the causes of contagion with the exhalations of an onion which induce lachrymation.

Appendix B

The 18th-century Preformationists, of whom Tristram Shandy
was one, were impressed with the idea that organisms contained in miniature within their bodies the germs or embryons
of all their descendants. The idea that there could be a limit
to this number was not discussed. Erasmus Darwin felt uneasy
about this view, sensing that it was incompatible with the
existence of atoms of a size that seemed to him reasonable. His
wording (1801) is interesting: '. . . That these infinitely minute
forms are only evolved or distended, as the embryon increases
in the womb. This idea, besides its being unsupported by any
analogy we are acquainted with, ascribes a greater tenuity to
organised matter than we can readily admit; as these included
embryos are supposed each of them to consist of the various
and complicate parts of animal bodies, they must possess a
much greater degree of minuteness than that which was ascribed to the devils that tempted St Anthony; of whom twenty
thousand were said to have been able to dance a saraband on
the point of the finest needle without incommoding each other.'

His grandson, Charles Darwin, was less convinced that
matter was not infinitely subdivisible. He records that he had
read the *Zoonomia*, 'but without producing any effect'. This
is clearly shown in his theory of 'pangenesis' by means of
'gemmules'. These 'gemmules' reproduced themselves, were
susceptible of modification which they then transmitted to
subsequent 'gemmules', and pervaded all parts of an organism. Darwin (1868) refers to this as a 'provisional hypothesis',
but it is set out in an unexpectedly declamatory and emotional
way with a rather casual use of words such as 'infinite' and

'inconceivable'. Nevertheless, it is interesting that at one point in the chapter he is troubled by the complexity of the inheritance mechanism he postulated. He wrote: 'Excessively minute and numerous as they are believed to be, an infinite number derived, during a long course of modification and descent, from each cell of each progenitor, could not be supported or nourished by the organism.' Charles, unlike Erasmus, did not realise that there simply would not be room for all the 'gemmules' he was postulating.

The difficulty was recognised, however, by Charles Darwin's cousin, the geneticist Francis Galton, who wrote succinctly in 1872: 'The heritage of peculiarities through the contributions of a thousand generations, even supposing a great deal of ancestral intermarriage, must far exceed what could be packed into a single ovum.' He suggested that only a representative group of traits must be transmitted, a suggestion which the physicist, J. Clerk Maxwell, treated, unfortunately for the progress of genetics at this time, with gentle derision. Maxwell, who calculated that micro-organisms or genes could contain a million molecules and that this would be enough for the transmission of all possible traits, preferred to emphasise the 'infinite complexity' of organisms. Galton sought a mechanism for heredity that could be set within the limits of the dimensions of a cell.

Notional biochemistry continued to flourish. Darwin's 'gemmules' were joined by Weismann's 'biophors', Tos's 'bionomes' and a host of other notional organules, the names of which are now forgotten. As with the more rococo developments of morphological evolution, a proliferation of terminology is a sign that a change is due. By the time of the publication of Perrin's paper on methods for the determination of molecular size (1909), it was apparent that Democritus and Dalton were right. Atoms had a real existence and a determinable size, as did the molecules which were the result of their preferred combining ratios. Research into the sizes of enzymes and other molecules of biological origin was on a sound basis (see Chapter 7) and the study of molecular biology (sometimes described as 'pop biochemistry') began.

Further Reading

For a general historical background to the subject:

HURD, D. L., AND KIPLING, J. J. (eds.) 1964. *The origins and growth of physical science*, Vol. II. Penguin Books, Harmondsworth, Middlesex.

OPARIN, A. I. 1957. *The origin of life on Earth*. Oliver & Boyd, Edinburgh.

ROOK, A. (ed.) 1964. *The origins and growth of biology*. Penguin Books, Harmondsworth, Middlesex.

Two useful sources of information on the origins of life:

FLORKIN, M. (ed.) 1960. *Aspects of the origin of life*. Pergamon Press, Oxford.

HALDANE, J. B. S. 1954. The origins of life. *New Biol.* **16**, 12.

Chapter 1

PIRIE, N. W. 1937. The meaninglessness of the terms 'life' and 'living'. In *Perspective in biochemistry*. Cambridge Univ. Press, London.

SCHAFER, E. A. 1912. Life, its maintenance, origin and nature. *Rep. Br. Ass. Advmt Sci.*, No. 74.

Chapter 2

ASHBY, W. R. 1952. Can a mechanical chess player outplay its designer? *Br. J. Phil. Sci.*, **3**, 44.

PENROSE, L. S. 1958. Mechanics of self-reproduction. *Ann. hum. Genet.*, **23**, 59.

Chapter 3

BULLOCH, W. 1938. *The history of bacteriology*. Oxford Univ. Press, London.

CRELLIN, J. K. 1966. The problem of heat resistance of micro-organisms. British Spontaneous Generation Controversies of 1860–80. *Med. Hist.*, **10**, 1.

HUXLEY, T. H. 1870. Biogenesis and abiogenesis. *Rep. Br. Ass. Advmt Sci.*, No. 73.

PASTEUR, L. 1860. Expériences relatives aux générations spontanées. *C. r. hebd. Séanc. Acad. Sci. Paris*, **348**, 52.

Chapter 4

BARGHOORN, E. S., AND TYLER, S. A. 1965. Micro-organisms of middle Pre-Cambrian Age from the Animikie Series, Ontario, Canada. In *Current aspects of exobiology*. Pergamon Press, Oxford.

GOLDSCHMIDT, V. M. 1952. Geochemical aspects of the origin of complex molecules on Earth. *New Biol.*, **12**, 97.

HALDANE, J. B. S. 1928. The origin of life. *Rationalist Ann.*, **3**. [Reprinted 1937 in collection of essays, *The inequality of man*. Penguin Books, Harmondsworth, Middlesex.]

PIRIE, N. W. 1953. Ideas and assumptions about the origins of life. *Discovery*, **14**, 238.

PRINGLE, J. W. S. 1954. The evolution of living matter. *New Biol.*, **16**, 58.

Chapter 5

PIRIE, N. W. 1964. The size of small organisms. Leeuwenhoek lecture. *Proc. R. Soc. B.*, **160**, 149.

PIRIE, N. W. 1970. Nature and origins of life. In DRONAMRAJU, K. R. (ed.) *Haldane and Modern Biology*. Johns Hopkins Press, Baltimore.

Chapter 6

BLOUT, E. R., AND IDELSON, M. 1956. Polymerisation of amino acid n-carboxyanhydrides. *J. Am. chem. Soc.*, **78**, 3857.

HALDANE, J. B. S. 1960. Pasteur and cosmic asymmetry. *Nature*, **185**, 87.

LATHAM, R. E. (trans.) 1950. (LUCRETIUS) *The nature of the universe*, Vol. II, line 485 ff. Penguin Books, Harmondsworth, Middlesex.

PIRIE, N. W. 1959. The position of stereoisomerism in the argument about the origins of life. *Trans. Bose Res. Inst.*, **22**, 111.

POWELL, H. M. 1952. The spontaneous optical resolution of solvated tri-*O*-thymotide. *J. chem. Soc.*, 3747.

Chapter 7

LWOFF, A. 1943. *L'évolution physiologique: études des pertes de fonctions chez les micro-organismes*. Hermann, Paris.

PIRIE, N. W. 1960. Chemical diversity and the origins of life. In FLORKIN, M. (ed.) *Aspects of the origin of life*. Pergamon Press, Oxford.

Name Index

Subject Index

Age of Earth, 36
Alcohol, 5, 6
Algae, 11
  fossil, 40
  vanadium as bio-element in, 77
Aluminium, 43, 71, 77
Amino acids, 46, 59, 68, 69, 70, 72, 73–76, 80, 84
Analogues, non-vital, 3, 11, 14–21, 44–47, 56
'Astroplankton', 30
Atmosphere
  Martial, 83
  terrestrial, 37–39, 73
    oxygenation of, 37–41, 47
  Venusian, 83

Bacillus subtilis, 29, Plate IV
Bacteria, x, 3, 4, 6, 7, 28, 29, 31–32, 41, 46, 49, 52, 55, 76, 80
Benzene, 65
Bible, The, 4, 22, 86
Biochemical evolution, x, 8, 40, 43–44, 55, 74–80
Biochemistry, 1
  anomalous aspects of present-day, 47, 76–80
Bio-elements, 76–79
Biological systems
  criteria for, 2–13, 45–47, 83–85
  replication of, ix, 2, 3, 6, 16–20, 23, 26
  structural organisation of, 21, 43, 45, 50–51, 56, 67
Biological uniformity, 48–51, 76

Biopoesis, 20, 24, 30, 32, 34, 41–47, 66, 73, 82
  possible location in atypical regions, 39, 64, 79
  relevance of simulated pro-biotic syntheses to, 73–74
Blood, 49, 55
British Association, Presidential Address, 1870, 33
    1911, 4, 69
Brownian movement, 1, 4, 20, 50, 81

Caesium, 79
Calcium, 43, 77
Cambrian period, 74
Carbohydrate, 8, 11, 16, 48
Carbon, 8, 10, 20, 27, 77
  'fossil', 37–39
  isotopes of, 10, 40
  quadrivalency of, 48
Carotenoids, 48
Catalysts, 6, 16–17, 46, 50, 81
  chemical models of biological, 45, 50, 63, 71
Cell, 6, 36, 50–52
  brain, 3
  membrane, 6, 48, 51–53, 56
  minimum size of, 3, 7, 35–48
  organelles, 48, 53, 55
Cellulose, digestion of, 7
Chert, Pre-Cambrian, 40, Plates V & VI
Chiral-directing agencies, 65–68
Chirality, 17 58–68
  at atomic level, 67–68
Cinchonine, 60

93